# The Mixing Engineer's Handbook

by Bobby Owsinski

*Edited by Malcolm O'Brien*

6400 Hollis Street
Emeryville, CA  94608

©1999 Intertec Publishing Corporation

Library of Congress Catalog Card Number: 99-62534


Cover Design: Linda Gough
Book Design and Layout: Linda Gough
Cover Photo: Susana Millman, Coverage Photography, San Francisco, CA
Back Cover Photo: Courtesy Marty Porter

Production Staff: Mike Lawson, publisher; Malcolm O'Brien, editor;
Sally Engelfried, editorial assistance

6400 Hollis St., Suite 12
Emeryville, CA 94608
(510) 653-3307

**Also from MixBooks:**

*The AudioPro Home Recording Course, Volumes I, II and III*
*I Hate the Man Who Runs this Bar!*
*How to Make Money Scoring Soundtracks and Jingles*
*The Art of Mixing: A Visual Guide to Recording, Engineering, and Production*
*500 Songwriting Ideas (For Brave and Passionate People)*
*Music Publishing: The Real Road to Music Business Success, Rev. and Exp. 4th Ed.*
*How to Run a Recording Session*
*Mix Reference Disc, Deluxe Ed.*
*The Songwriters Guide to Collaboration, Rev. and Exp. 2nd Ed.*
*Critical Listening and Auditory Perception*
*Keyfax Omnibus Edition*
*Modular Digital Multitracks: The Power User's Guide*
*The Dictionary of Music Business Terms*
*Professional Microphone Techniques*
*Concert Sound*
*Sound for Picture*
*Music Producers*
*Live Sound Reinforcement*

**Also from EMBooks:**

The Independent Working Musician
Making the Ultimate Demo
Anatomy of a Home Studio
The EM Guide to the Roland VS-880

MixBooks is a property of Intertec Publishing Corporation
Printed in Auburn Hills, MI
ISBN 0-87288-723-5

# Contents

# Foreword

I wrote this book probably for the same reason that you're reading it; that is, to get better at what I do. I noticed over the years that my mixes were somewhat hit or miss. Sometimes they were great, sometimes OK, sometimes just plain off the mark. I also noticed that much of the time my mixes didn't have the "big time" sound that I heard on the radio. I wanted this sound badly and the only way I knew how to get it was to ask questions of the engineers that already knew the secret.

While doing research for this book I found that a common factor among most good mixers was that, for the most part, they all had at least one mentor as a result of coming up through the studio ranks. Most great mixers started as assistants and learned by watching and listening to the greats that they helped and had taken a little from all of them as a result.

I hadn't done that, however. Being a musician first and foremost, I learned to engineer thanks to my early interests in electronics in general and how the electrons got from my guitar to the speakers of my amplifier specifically. As I became familiar with the recording studio, I was quickly offered all sorts of session work, from recording jingles to big band to Jazz to R&B to Hard Rock. But, never wanting to give up being a musician (which I knew I'd have to do), I never took a proper studio job to really learn the trade at the hands of the masters. As a result, my recording skills were always pretty good, but my mixing skills were lacking.

Having taught recording for many years at Berklee College of Music, Trebas Recording Institute, and Nova Institute Multimedia Studies, I soon realized that there were many others like me who were good, but not great. It wasn't that they weren't capable, but they didn't have the opportunity or access to the methods of the masters. After all, how often does a George Massenburg or Bruce Swedien record in Lincoln, Peoria or Santa Fe?

In almost all cases, the mixers that I talked to were extremely forthcoming, answering just about any question and offering explicit information as to why and how they work. Professional jealousy just does not exist in this industry; at least, not in my experience.

So the book started out very selfishly, as it was meant specifically to meet my needs, but it ended up for you as well. I hope you will benefit from it as much as I have.

And yes, my mixes have gotten much, much better.

# *Preface*

This book changed direction several times as it was being written. Since I had met a lot of top engineers over the years, I originally thought that I'd ask them how they mixed purely as background material (a reference, if you will), and simply accumulate their various methods and anecdotes.

The more I got into it though, the more it became obvious that these interviews were living and breathing on their own and they really should be included in their entirety in the text. Otherwise, a lot of really useful information would be left out. In other words, I let them tell you what they do in their own words.

## MEET THE MIXERS

By way of introduction, here's a list of the engineers who contributed to this book along with some of their credits. I've tried to include someone that represents every genre of modern music (Punk to Classic to Alternative to Jazz to Classical to R&B to Dance to Re-mixing to Latin to Rap to Orchestral to Country to TV mixing) so there's something for everyone. I'll be quoting them from time to time so I wanted to introduce them early on so you have an idea of their background when they pop up.

Just remember that whenever a "mixer" or "engineer" is referred to in this book, I don't mean your average run of the mill Joe Blow engineer (hard working and well meaning though he may be). I mean someone who's made the hits that you've heard and loved and whose sounds you've tried to imitate. This book is about how these glorious few think, how they work, and why they do the things they do. And even though we can't hear as they hear, perhaps we can hear through their words.

**Joe Chiccarelli**  He may not have quite as high a profile as many other notable big-time mixers, but engineer/producer Joe Chiccarelli's list of projects are as notable as the best of the best. With credits like Tori Amos, Etta James, Beck, U2, Oingo Boingo, Shawn Colvin, Frank Zappa, Bob Seger, Brian Setzer, Hole and many, many more, chances are you've heard Joe's work more times than you know.

**Lee DeCarlo** From his days as chief engineer at LA's Record Plant in the heady 70's, Lee DeCarlo has put his definitive stamp on hit records from Aerosmith to John Lennon's famous *Double Fantasy* to current releases by Rancid and Zak Wylde.

**Benny Faconne** Benny is unique in that he's a Canadian (from Montreal), but 99 percent of the things that he works on are Spanish. From five Luis Miguel records to Ricky Martin to the Latin rock band Mana to the Spanish re-mixes for Boys 2 Men, Tony Braxton and Sting, Benny's work is heard far and wide around the Latin world.

**Jerry Finn** With credits like Green Day to Rancid to the Goo Goo Dolls to Beck, Jerry represents one of the new generation of mixers who knows all the rules but is perfectly willing to break them.

**Jon Gass** Babyface, Tony Rich, Mariah Carey, Usher, the *Waiting To Exhale* soundtrack. Mixer Jon Gass' credit list reads like a *Who's Who* of R&B greats and with good reason. Gass' unsurpassed style and technique has elevated him to a most esteemed position among engineers, working with the best of the best on some of the most creative and demanding music around today.

**Don Hahn** Although there's a lot of pretty good engineers around these days, not many have the ability to record a 45- to 100-piece orchestra with the ease of someone who's done it a thousand times. Don Hahn can, and that's because he actually *has* done it a thousand times. With an unbelievable list of credits that range from television series like *Star Trek (The Next Generation, Deep Space Nine* and *Voyager), Family Ties, Cheers* and *Columbo* to such legends as Count Basie, Barbara Streisand, Chet Atkins, Frank Sinatra, Herb Alpert, Woody Herman, Dionne Warwick and a host of others (actually ten pages more), Don has recorded the best of the best.

**Ken Hahn** There are few people that know TV sound the way Ken Hahn does. From the beginning of the television post revolution, Hahn's New York-based Sync Sound has lead the way in television sound innovation and the industry's entry into the digital world. Along the way Ken has mixed everything from *PeeWee's Playhouse* to concerts by Billy Joel and Pearl Jam and a host of others while picking up a slew of awards in the process (four Emmys, a CAS award and 13 ITS Monitor awards).

**Andy Johns** Andy Johns needs no introduction because we've been listening to the music that he's mixed for most of our lives. With credits like Led Zeppelin, Free, Traffic, Blind Faith, The Rolling Stones and most recently Van Halen (to name just a few), Andy has set a standard that most mixers are still trying to achieve.

**Kevin Killen** From Peter Gabriel's seminal *So* to records by U2, Elvis Costello, Stevie Nicks, Bryan Ferry and Patty Smith (to name just a few), Kevin Killen's cutting edge work has widely influenced an entire generation of mixers.

**Bernie Kirsh** Bernie Kirsh has certainly made his mark as one of the top engineers in the world of Jazz. From virtually all of Chick Corea's records to working on Quincy Jones' ground breaking *Back on the Block* (which won a Best Engineering Grammy), Bernie's recordings have consistently maintained a level of excellence that few can match.

**George Massenburg** From designing the industry's most heralded audio tools to engineering classics by Little Feat, Earth, Wind and Fire and Linda Ronstadt (to name only a few), George Massenburg needs no introduction to anyone even remotely connected to the music or audio business.

**David Pensado** With projects by Bel Biv Devoe, Coolio, Take 6, Brian McKnight, Diana Ross, Tony Toni Tone, Atlantic Starr and many more, David Pensado has consistently supplied mixes that have not only filled the airwaves, but ranked among the most artful as well.

**Ed Seay** Getting his start in Atlanta in the 70's by engineering and producing hits for Paul Davis, Peabo Byson and Melissa Manchester, Ed Seay has since become one of the most respected engineers in Nashville since moving there in 1984. With hit-making clients such as Pam Tillis, Highway 101, Collin Raye, Martina McBride, Ricky Skaggs and a host of others, Ed has led the charge in changing the way that recording is approached in Nashville.

**Allen Sides**   Although well known as the owner of the premier Oceanway studio complexes in both Los Angeles and Nashville, Allen is one of the most respected engineers in the business. His recent credits include the film scores to *Dead Man Walking, Phenomenon, Last Man Standing* and Michael Cimino's *Sunchaser,* as well as records with the Brian Setzer Big Band and *The Songs of West Side Story,* featuring Phil Collins. His credits also include Natalie Cole, All For One, Trisha Yearwood, Wynonna Judd, Tevin Campbell, Kenny Loggins, Michael McDonald, Little Richard and Aretha Franklin.

**Don Smith**   With credits that read like a *Who's Who* of Rock & Roll, Don has lent his unique expertise to projects by The Rolling Stones, Tom Petty, U2, Stevie Nicks, Bob Dylan, Talking Heads, The Eurythmics, The Traveling Wilburys, Roy Orbison, Iggy Pop, Keith Richards, Cracker, John Hiatt, The Pointer Sisters, Bonnie Raitt and lots more.

**Guy Snider**   A former guitar player who played with the likes of Ike and Tina Turner and Chuck Berry, Guy Snider's projects as an engineer run the musical gamut from rockers The Rolling Stones, Nine-Inch Nails and Faith No More to the smooth R&B of Brandy and Shante Moore to hard core rappers Tupac, Snoop Doggy Dog and Nate Dog.

**Ed Stasium**   Producer/engineer Ed Stasium is widely known for working on some of the best guitar albums in recent memory, including my own personal favorites by the Smithereens, Living Color and Mick Jagger. From Marshall Crenshaw to Talking Heads to Soul Asylum to Motorhead to Julian Cope to the Ramones to even Gladys Knight and the Pips and Ben Vereen, Ed has put his indelible stamp on their records as only he can.

**David Sussman**   While engineering for noted dance music producers David Morales and Grammy Award-winning Re-mixer of the Year Frankie Knuckles, David has developed quite a resume, engineering re-mix work for such artists as Mariah Carey, Whitney Houston, Janet Jackson, Madonna, Tina Turner, Gloria Estefan, Seal, Michael Jackson and U2, as well as recent additional production and mix credits for Salt-N-Pepa and MLF.

**Bruce Swedien**  Maybe the most revered of all mixers, Bruce has a credit list (as well as five Grammys) that could take up a chapter of this book alone. Although Michael Jackson would be enough for most mixers' resumes, Bruce can also include such legends Count Basie, Tommy Dorsey, Duke Ellington, Woody Herman, Oscar Peterson, Nat "King" Cole, George Benson, Mick Jagger, Paul McCartney, Patti Austin, Edgar Winter and Jackie Wilson among many, many others.

**John X**  John X. Volaitis is one of the new breed of engineers who's thrown off his old school chains and ventured into the world of re-mixes. Along with his partner Danny Saber, John has done recent re-mixes for such legends as David Bowie (Dead Man Walking), (Little Wonder) and U2 (Staring at the Sun), as well as Marilyn Manson (Horrible People), Garbage (Stupid Girl) and a host of others.

---

For those of you who don't have the time or desire to read each interview, I've summarized many of the working methods that these great engineers use in *Part I — Mixing in Stereo*.

Part Two is exclusively about mixing in surround, which is an ability that will be needed by every mixer in the near future. This is something about which I do know a little, having been one of the early devotees of the 5.1 format. Much of the info in this section comes from personal mixing and recording experience and techniques that I've gathered as an editor for *Surround Professional* magazine.

**DISCLAIMER:**  **Just because you read this book doesn't automatically guarantee that you'll become a great platinum mixer who makes lots of money and works with big recording artists. You'll get some tips, techniques and tricks from the book but you still need ears and experience, which only you can provide. All this book can do is point you in the right direction and help a little on the way!**

Keep in mind that just because a number of mixers do things a certain way, that doesn't mean that's the only way to do it. You should always feel free to try something else, because after all, whatever works for you is, in fact, the right way.

# Part I
## Mixing in Stereo

# CHAPTER 1

# The Evolution of Mixing

 efore we get into the actual mechanics of mixing, it's important to have some perspective on how this art has developed over the years.

It's obvious to just about everyone who's been around long enough that mixing has changed over the decades, but the why's and how's aren't quite so obvious. In the early days of recording in the 50's, there really wasn't any mixing per se since the recording medium was mono and a big date used only four microphones. Of course, over the years recording developed from capturing an unaltered musical event to one that was artificially created through overdubs, thanks to the innovation of Selsync (the ability to play back off of the record head so everything stayed in sync) introduced in 1955. The availability of more and more tracks begat larger and larger consoles, which begat computer automation and recall just to manage the larger consoles fed by more tracks. With all that came not only an inevitable change in the philosophy of mixing but a change in the way that a mixer listened or thought as well.

According to the revered engineer/producer Eddie Kramer, "Everything (when I started recording) was 4-track, so we approached recording from a much different perspective than people do nowadays. My training in England was fortunately with some of the greatest engineers of the day, who were basically classically trained in the sense that they could go out and record a symphony orchestra and then come back to the studio and then do the Jazz or Pop, which is exactly what we used to do. When I was training under Bob Auger, who was the senior engineer at Pye Studios, he and I used to go out and do classical albums with a 3-track Ampex machine and three Neumann U47's and a single mixer of three channels. So with that sort of training and technique under my belt, approaching a Rock & Roll session was approaching it from a classical engineering standpoint and making the sound of a rock band bigger and better than it was. But the fact of the matter was that we had very few tools at our disposal except EQ, compression, and tape delay. That was it."

English mixer Andy Johns, who apprenticed under Kramer and eventually went on to equally impressive credits with The Rolling Stones, Led Zeppelin, Traffic, Van Halen and others, goes a step further. "You know why the Beatles' *Sgt. Pepper's* sounds so good? You know why *Are You Experienced?* sounds so good, almost better than what we can do now? Because, when you were doing the 4-to-4 (bouncing down from one 4-track machine to another), you mixed as you went along. There was a mix on two tracks of the second 4-track machine and you filled up the open tracks and did the same thing again. Listen to "We Love You" (by the Stones). Listen to *Sergeant Pepper's*. Listen to *Hole in My Shoe* by Traffic. You mixed as you went along. Therefore, after you got the sounds that would fit with each other, all you had to do was adjust the melodies. Nowadays, because you have this luxury of the computer and virtually as many tracks as you want, you don't think that way any more."

And indeed, once more tracks were available and things began to be recorded in stereo, the emphasis turned from the bass anchoring the record to the big beat of the drums as the main focal point. This is partially because drum miking typically went from just overhead and kick drum mics to the now common occurrence of a mic on every drum, since the consoles could now accommodate more microphone inputs and there were plenty of tracks on which to record. And, since the drums could be spread out over six or eight or even more tracks, they could be concentrated on more during the mix because they didn't have to be pre-mixed along with the bass onto only one or two tracks. Instead of the drums being thought of as just another instrument equal to the bass, they now demanded more attention because more tracks were used.

At that point (approximately 1975), thanks to the widespread use of the now standard 24-track tape deck, mixing changed forever. And, for better or for worse, mixing changed into what it is today.

Although there's less of a distinction these days than there used to be, where you live has a great influence on the sound of your mix. Up until the late 80's or so, it was easy to tell where a record was made, just by its sound. There's been a homogenization of styles in recent years, mostly because engineers now mix in a variety of locations and many have relocated to new areas, transplanting their mixing styles along the way.

There are three major recording styles and most recordings fall into one of them; New York, LA and London.

### The New York Style

The New York style is perhaps the easiest to identify because it features a lot of compression, which makes the mix very punchy and aggressive (just like New Yorkers). In many cases, the compressed instruments (mostly the rhythm section) are even recompressed several times along the way. It seems that every New York engineer that I interviewed (even the transplanted ones) had virtually the same trick. Send the drums (sometimes with the bass) into a couple of busses, send that through some compressors, squeeze to taste, then add back a judicious amount of this compressed rhythm section to the mix through a couple of faders. This can be enhanced even further by boosting the high and low frequencies (lots of boost in most cases) to the compressed signal as well (more on this New York Compression Trick later in the book in the chapter on Dynamics). For an example of this, listen to any of the mixes that Ed Stasium (a proud practitioner of this method) has done, such as the Mick Jagger solo album *She's the Boss*, or anything by The Smithereens or Living Color.

### The LA Style

The LA sound is a somewhat more natural sound; it is compressed, but to a less obvious degree than the New York style. There's also a lot less effects layering than the London style. The LA style has always tried to capture a musical event and augment it a little, rather than recreate it. Some good examples would be any of the Doobie Brothers or Van Halen hits of the 70's and 80's.

**The London Style**

The London sound is a highly layered musical event that borrows some from the New York style in that it's somewhat compressed, but deals with multiple effect layers. This style makes extensive use of what is known as *perspective*, which puts each instrument into its own distinct sonic environment. Although musical arrangement is important to any good mix, it's even more of a distinctive characteristic of a London mix. What this means is that many parts appear at different times during a mix; some for effect, some to change the dynamics of the song. Each new part will be in its own environment and as a result will have a different perspective. A perfect example of this would be Yes' *Owner of a Lonely Heart* or just about anything done by Trevor Horn, such as Seal or Grace Jones.

As we approach the millennium, there's much less of a difference between styles than there was during the 80's, but variations still do exist. Although the style differences blur on most music, Techno and Dance still have considerable variation divided around the traditional geographic boundaries of London, New York and LA

**Other Styles**

Increased globalization has had its effect on regional styles, too. Where once upon a time Philadelphia, Memphis, Ohio, Miami and San Francisco all had sub-styles of the Big Three, all these areas now line up clearly in one of the Big Three camps.

Nashville today is a special case among the regional styles, though. This is a style that's evolved (some might say devolved) from an offshoot of the New York style during the 60's and 70's to become much more like the LA sound of the 70's. Says engineer/producer Ed Seay, "Back when I used to listen to my dad's old Ray Price and Jim Reeves Country records, they weren't very far from what Pop was in the early 60's. Very mellow, big vocals, very subdued band, very little drums, strings, horns, lush. Mix-wise, there wasn't really too much difference in an Andy Williams record and one of the old Jim Reeves records.

"What happened was that Country got too soft sounding. You'd cut your track and then do some sweetening with some horns and strings. At one time, strings were on all the Country records and then it kind of transformed into where it's at today, with almost no strings on Country records, except for big ballads. For the most part, horns are completely dead. They're almost taboo. Basically, it's rhythm track-driven and not really very far off from where Pop was in the mid-to-later 70's. The Ronstadt "It's So Easy To Fall In Love" and "You're No Good," where you hear guitar, bass, drums, keyboards, a slide or steel and then a vocal background; that's pretty much the format now, although fiddle is used also. Ironically enough, a lot of those guys that were making those records have moved here because at this point, this is one of the last bastions of live recording."

The globetrotting lifestyle of the 90's engineer has caused a homogenization of regional styles. At one time, most studios had house engineers; today the market is predominately made up of freelancers that freely travel from studio to studio, project to project, bouncing between different cities (and therefore styles) as easily as flipping the channel on a TV. At one time, an engineer might change studios but remain located in a specific area all his working life; now it's not uncommon for an engineer to relocate to several major media centers during the course of his career. All this means a cross-pollination of styles, which blurs the distinction between the Big Three as we move into the next millenium.

# The Mechanics of Mixing

 lthough most engineers ultimately rely upon their intuition when doing a mix, there are certain mixing procedures that they all consciously or unconsciously follow.

## HEARING THE FINAL PRODUCT

By and large, most mixers can hear some version of the final product in their heads before they even begin to mix. Sometimes this is a result of countless rough mixes during the course of a project, which gradually become polished thanks to console automation and computer recall, if an engineer is mixing a project that he's tracked. Even if an engineer is brought in specifically to mix, many won't even begin until they have an idea of where they're going.

Engineers who can hear the finished product before they start normally begin a mix the same way. They become familiar with the song either through a previous rough mix or by simply putting up all the faders and listening for a few passes. Sometimes this is harder than it seems, though. In the case of a complex mix with a lot of track-sharing or synced multi-tracks, the mix engineer may have to spend some time writing mutes (a *cut pass*) before the song begins to pare down and make sense.

**Ed Seay:**  *I think one of the things that helps me as a mixer, and one thing that helps all of the ones that have made a mark, is what I call "having the vision." I always try to have a vision of the mix when I start. Rather than just randomly pushing up faders and saying, "Well, a little of this EQ or effect might be nice," I like to have a vision as far as where we're going and what's the perspective.*

For better or worse, the engineer's vision will end up changing, thanks to input from the producer and/or artist in most cases. Although sometimes a major mixer will complete the job unattended by the producer/artists, most mixers actually prefer the input. However, the vast majority would prefer to start the mix by themselves and have the artist at hand to offer suggestions five or six hours later, after the mix begins to take shape.

## THE OVERALL APPROACH

Whether they know it or not (and many mixers aren't conscious of how they do it), most great mixers have a method in the way they approach a mix. Although the method can vary a little depending on the song, the artist, and the genre or if the mixer tracked the song from scratch or is just coming in for the mix, the technique remains constant.

> **Figure out the direction of the song.**
> **Develop the groove and build it like a house.**
> **Find the most important element and emphasize it.**

The last point may be the most important in creating an outstanding mix. As famed Latin mixer Benny Faconne so succinctly states, "It's almost like a musician who picks up a guitar and tries to play. He may have the chart in front of him, but soon he has to go beyond the notes in order to get creative. Same thing with mixing. It's not just a thing of setting levels any more, but more about trying to get the energy of the song across. Anybody can make the bass or the drums even out."

## TALL, DEEP AND WIDE

Most great mixers think in three dimensions. They think, "Tall, deep and wide," which means to make sure that all the frequencies are represented; make sure there's depth to the mix, then give it some stereo dimension as well.

The "tall" dimension (which is called Frequency Range later in the book) is the result of knowing what sounds right, due to having a reference point. This reference point can come from being an assistant engineer and listening to what other first engineers do, or simply by comparing your mix to some CDs, records or tapes that you know and consider to be of high fidelity.

Essentially, what you're trying to accomplish is to make sure that all the frequencies are properly represented. Usually that means that all of the sparkly, tinkly highs and fat, powerful lows are there. Sometimes some mids need to be cut. Clarity is your goal. Again, experience with good sounds really helps as a reference point.

The effects or "deep" dimension is achieved by introducing new ambience elements into the mix. This is usually done with reverbs and delays (and offshoots like flanging and chorusing), but room mics, overheads and even leakage play an equally big part as well.

The panning or "wide" dimension is placing a sound element in a sound field so as to make a more interesting soundscape, such that each element is heard more clearly.

Which brings us to the nitty-gritty of the book, where all the elements of a great mix are detailed even further.

## THE SIX ELEMENTS OF A MIX

Every piece of modern music—meaning Rock, Pop, R&B, Rap, Country, AOR, CHR, New Age, Swing, and every other genre having a strong back-beat—has six main elements to a great mix.

They are:
**Balance** — the volume level relationship between musical elements
**Frequency Range** — having all frequencies properly represented
**Panorama** — placing a musical element in the sound field
**Dimension** — adding ambience to a musical element
**Dynamics** — controlling the volume envelopes of a track or instrument
**Interest** — making the mix special

Many mixers have only four or five of these when doing a mix, but all of these elements MUST be present for a GREAT mix or a hit mix, as they are all equally important.

In music requiring simple recreation of an unaltered acoustic event (Classical or Jazz or any live concert recording), it's possible that only the first four elements are needed to have a mix be considered great. Dynamics and Interest have evolved to become extremely important elements as modern music has evolved.

We'll look at each element individually in the coming chapters.

# Element One: Balance — The Mixing Part of Mixing

T he most basic element of a mix is *balance*. A great mix must start here first; for without balance the other mix elements pale in importance. There's more to balance than just moving some faders, though, as we'll see.

## THE ARRANGEMENT — WHERE IT ALL BEGINS

Good balance starts with good arrangement. It's important to understand arrangement since so much of mixing is subtractive by nature. This means that the arrangement (and therefore the balance) is changed by the simple act of muting an instrument that doesn't fit well with another. If the instruments fit well together and don't fight one another, the mixer's life becomes immensely easier. But what exactly does "fighting one another," mean?

When two instruments with essentially the same frequency band play at the same volume at the same time, the result is a fight for attention. Think of it this way: you don't usually hear a lead vocal and a guitar solo at the same time, do you? That's because the listener is unable to focus on both simultaneously and becomes confused and fatigued as a result.

So how do you get around instrument "fighting"? First and foremost is a well-written arrangement, which keeps instruments out of each other's way right from the beginning. The best writers and arrangers have an innate feel for what will work in an arrangement, and the result is one that lies together without much help, almost automatically.

But it's not uncommon to work with an artist or band that isn't sure of the arrangement or is into experimenting and just allows an instrument to play throughout the entire song, thereby creating numerous conflicts. This is where the mixer gets a chance to rearrange the track by keeping what works and muting the conflicting instrument or instruments. Not only can the mixer influence the arrangement this way, but also the dynamics and general development of the song.

In order to understand how arrangement influences balance, we have to understand the mechanics of a well-written arrangement first.

Most well conceived arrangements are limited in the number of elements that occur at the same time. An element can be a single instrument like a lead guitar or a vocal, or it can be a group of instruments like the bass and drums, a doubled guitar line, a group of backing vocals, etc. *Generally, a group of instruments playing exactly the same rhythm is considered an element.* Examples: a doubled lead guitar or doubled vocal is a single element, as is a lead vocal with two additional harmonies. Two lead guitars playing different parts are two elements, however. A lead and a rhythm guitar are two separate elements as well.

## ARRANGEMENT ELEMENTS

**Foundation** — The rhythm section. The foundation is usually the bass and drums, but it can also include a rhythm guitar and/or keys if they're playing the same rhythmic figure as the rhythm section. Occasionally, as in the case of power trios, the foundation element will only consist of drums, since the bass will play a different rhythm figure and become its own element.

**Pad** — A pad is a long sustaining note or chord. In the days before synthesizers, a Hammond Organ provided the best pad and was joined later by the Fender Rhodes. Synthesizers now provide the majority of pads, but real strings or a guitar power chord can also suffice.

**Rhythm** — Rhythm is any instrument that plays counter to the foundation element. This can be a double-time shaker or tambourine, a rhythm guitar strumming on the back-beat or congas playing a Latin feel. The rhythm element is used to add motion and excitement to the track.

**Lead** — A lead vocal, lead instrument or solo.

**Fills** — Fills generally occur in the spaces between lead lines, or they can be a signature line. You can think of a fill element as an answer to the lead.

That's not to say that each individual instrument is a separate element, however. In Bob Seger's hit "Night Moves," there are bass and drums, acoustic guitar, piano, Hammond organ, lead vocal and background vocals. This is how they break out:

**Bob Seger's "Night Moves"**

> **Foundation** — Bass, drums, acoustic guitar
>
> **Pad** — Hammond organ
>
> **Rhythm** — Piano
>
> **Lead** — Lead vocal
>
> **Fills** — Background vocal answers and sometime the piano fills in the holes

Usually an acoustic guitar falls into the rhythm category as the strumming is pushing the band and creating excitement. In "Night Moves," however, the acoustic guitar is pulled back level-wise in the mix so it melds into the rhythm section, effectively becoming part of the foundation element.

Alanis Morissette's "Thank U" contained several good examples of both rhythm and pads. What's different is that there are two sets of each, one for the intro and chorus, and a different set for the verses.

**Alanis Morissette's "Thank U"**

> **Foundation** — Bass, drums
>
> **Pad** — Synthesizer in intro and chorus behind the piano; different synths in chorus
>
> **Rhythm** — Piano; "breath" sample in the verse
>
> **Lead** — Lead vocal
>
> **Fills** — Guitar fills in the second verse

Of course, there's much more going on in this song track-wise, but any additional tracks are either replacing or doubling the above elements. The number of elements remains constant.

**Garth Brook's "Two Pina Coladas"**

> **Foundation** — Bass, drums
>
> **Pad** — Steel guitar
>
> **Rhythm** — Acoustic guitar and shaker
>
> **Lead** — Lead vocal
>
> **Fills** — Electric and acoustic lead guitar; occasional steel fill

This song is different because there's no true pad in the traditional sense; but the steel guitar playing softly in the background acts the part well and shows that it's possible for non-traditional instruments to play that role.

## RULES FOR ARRANGEMENTS

There are a couple of easy-to-remember rules that will always make even the densest arrangement manageable.

### Limit the Number of Elements

Usually there should not be more than four elements playing at the same time. Sometimes three elements can work very well. Very rarely will five elements simultaneously work.

**Kevin Killen:** *I had an experience about three years ago on a Stevie Nicks record with Glyn Johns, who's been making records since the 50's. We were mixing without automation and he would just push the faders up and within a minute or two he would have this great mix. Then he would just say that he didn't like it and pull it back down again and push it back up. I relearned that the great art of mixing is the fact that the track will gel almost by itself if it was well performed and reasonably well recorded. I find that the stuff that you really have to work a lot harder on is the stuff that has been isolated and really worked on. The tracks all end up sounding like disparate elements and you have to find a way to make them blend together.*

**Everything in Its Own Frequency Range**

The arrangement (and therefore the mix) will fit together better if all instruments sit in their own frequency range. For instance, if a synthesizer and rhythm guitar play the same thing in the same octave, they will usually clash. The solution would be to change the sound of one of the instruments so they fill different frequency ranges—have one play in a different octave, or have them play at different times but not together.

**Lee DeCarlo:** *So much of mixing is what you take away, either level-wise or frequency-wise. There are so many things that you have to eliminate in order to make it all sit and work together. Mark Twain once said, "Wagner's music is much better than it sounds." Wagner is a guy that wrote for cellos and French horns doing things in the same register, but it all worked. The only reason that it worked was he kept the other things out of their way. If you have an orchestra and everybody's playing in the same register, its just going to get away on you. But if you leave holes, then you can fill up the spectrum.*

**Figure 1**

## Ways to Prevent Instrument Fighting

• Change the arrangement and rerecord the track

• Mute the offending instruments so that they never play at the same time

• Lower the level of the offending instrument

• Tailor the EQ so that the offending instrument takes up a different frequency space

• Pan the offending instrument to a different location

## WHERE TO BUILD THE MIX FROM

Different mixers start from different places when building their mix. This has as much to do with training as it does with the type of material. For instance, most old-time New York mixers and their proteges usually start from the bass guitar and build the mix around it. Many other mixers work from the drum overheads first, tucking in the other drums as they go along. Many mixers mix with everything up, only soloing specific instruments that seem to exhibit a problem. Still others are completely arbitrary, changing the starting place from song to song depending upon whatever instrument needs to be focused on.

**Joe Chiccarelli:** *Usually what I do is put up all the faders first and get a pretty flat balance and try to hear it like a song, then make determinations from there whether to touch up what I have or rip it down and start again from the bottom.*

**Jon Gass:** *I start with everything on and I work on it like that. The reason is that, in my opinion, the vocal is going to be there sooner or later anyway. You might as well know where it's sitting and what it's doing. All the instruments are going to be there sooner or later so you might as well just get used to it. And I think that's also what helps me see what I need to do within the first passage.*

**John X:** *I generally have to start with the loops. You've got to find the main loop or the combination of loops that creates the main groove. Sometimes the loops may have a lot of individual drums, but they're usually not crucial rhythmic elements. They can be accents and they can be stuff that just pops up in a break here and there.*

**Ken Hahn:** *It's usually vocals again. I make sure that those are perfect so that it becomes an element that you can add things around. I always clean up the tracks as much as I can because inevitably you want to get rid of rumble and thumps and noises, creaks, mic hits, etc. Then I always start with bass and rhythm.*

**Benny Faccone:** *It really is like building a house. You've got to get the foundation of bass and drums and then whatever the most important part of the song is, like the vocalist, and you've got to build around that. I put the bass up first, almost like the foundation part. Then the kick in combination with the bass to get the bottom. Because sometimes you can have a really thin kick by itself, but when you put the bass with it, it seems to have enough bottom because the bass has more bottom end. I build the drums on top of that. After I do the bass and drums, then I get the vocal up and then build everything from there. A lot of mixers just put the music up first, but as soon as you put the vocal up, the levels become totally different. After all the elements are in, I spend maybe a couple of hours just listening to the song like an average listener would, and I keep making improvements.*

**Ed Seay:** *I'll usually go through and push up instruments to see if there are any trouble spots. All this is dependent upon whether it's something that I've recorded or if I'm hearing it fresh and have no idea what it is. If that's the case, then what I'll do is rough-mix it out real quick. I'll push it up and see where it's going before I start diving in.*

*If it's something that I know what's on the tape, then I'll go through and mold the sounds in a minor way to fit the modern profile that it needs to be. In other words, if it's a real flabby, dull kick drum, it doesn't matter what the vision is. This kick drum's never going to get there. So I'll pop it into a Vocal Stresser or I'll do whatever I have to do. I'll work through my mix like that and try to get it up into the acceptable range, or the exceptional range, or at least somewhere that can be worked with. It takes a couple of hours to get good sounds on everything and then another couple of hours to get real good balances, or something that plays itself as if it makes sense. Then I'll do some frequency juggling so that everybody is out of everybody else's way.*

Wherever your starting point may be, it's generally agreed that the vocal (or whatever is the most prominent or significant melody instrument) has to make its entrance into the mix as soon as possible. The reason for this is two-fold. First of all, the vocal is probably going to be the most important element, so it will take up more frequency space than other supporting instruments. If you wait until late in the mix to put in the vocal, there may not be enough space left and the vocal will never sit right with the rest of the track.

The second reason has to do with effects. If you tailor all your effects to the rhythm section and supporting instruments, there may be none left when it's time to add in the vocal or most prominent instrument.

Figure 2

## Typical Mix Starting Places

- From the bass
- From the kick drum
- From the snare drum
- From the overheads
- From the lead vocal or main instrument
- When mixing a string section, from the highest string (violin) to the lowest (bass)

## WHAT TYPE OF PROGRAM MATERIAL?

The type of program being mixed will frequently have an effect on where you start building the mix. For instance, when doing Dance music where the kick is everything, that is the obvious choice for a starting point. When mixing something orchestral however, the emphasis is different. According to Don Hahn, "The approach is totally different because there's no rhythm section. So you shoot for a nice roomy orchestral sound and get as big a sound as you can get with the amount of musicians you have. You start with violins, then violas if you have them, cellos, then basses. You get all that happening and then add woodwinds, French horns, trombones, trumpets and then percussion and synthesizers if needed."

In Jazz, the melody will be the starting point with the bass inserted afterward to solidify the foundation.

## LEVEL-SETTING METHODS

Setting levels by using the VU meters has been debated from the beginning of mixing time. Some mixers feel that they can get in the ballpark by setting the levels with the meters alone while others discount any such method out of hand. The fact of the matter is that for those using the meter method, feel and instinct are still a large part of their technique, making it equally as valid as those who rely solely on instinct.

As with everything else that you read, try the following methods, use what works and throw away the rest.

**Benny Faccone:** *I usually start with the bass at about -5 and the kick at about -5. The combination of the two, if it's right, should hit about -3 or so. By the time the whole song gets put together and I've used the computer to adjust levels, I've trimmed everything back somewhat. The bass could be hitting -7 if I solo it after it's all done.*

**Don Smith:** *I'll start out with the kick and bass in that area (-7VU). By the time you put everything else in it's the total mix +3 anyway. At least if you start that low you have room to go.*

**Ed Seay:** *Usually a good place to start is the kick drum at -6 or -7 or so. I'll try to get a bass level that is comparable to that. If it's not exactly comparable on the meter because one's peaking and one's sustaining, I get them to at least sound comparable because later, in mastering, if you affect one, you're going to affect the other. So as long as the ratio is pretty correct between the two, then if you go to adjust the kick at least it's not going to whack the bass way out as long as they relate together. That's kind of a good starting place for me.*

**Lee DeCarlo:** *I'll get the snare drum constantly hitting the back-beat of the tune at around -5, and everything gets built around it.*

# CHAPTER 4

# Element Two: Panorama — Placing the Sound in the Soundfield

T One of the most overlooked or taken for granted elements in mixing is *panorama,* or the placement of a sound element in the sound field. To understand panorama, first we must understand that the stereo sound system (which is two channels for our purposes) represents sound spatially. Panning lets us select where in that space we place the sound.

In fact, panning does more than just that. Panning can create excitement by adding movement to the track and adding clarity to an instrument by moving it out of the way of other sounds that may be clashing with it. Correct panning for a track can also make it sound bigger, wider and deeper.

So what is the proper way of panning? Are there any rules? Well, like so many other things in mixing, although panning decisions may sometimes seem arbitrary, there's a method to follow and a rationale behind the method as well.

Imagine that you're at the movies and watching a Western. The scene is a panorama of the Arizona desert and right in the middle of the screen is a cowboy sitting on his horse in a medium shot from his boots up. Now a pack of Indians (we'll say six) are attacking him but we can't see them because the cowboy is in the shot directly in front of them. If we can't see them, their impact as a suspense builder is really limited, not to mention the fact that they cost the production money which just went to waste. Wouldn't it be better if the director moved the Indians to the left out of the shadow of the cowboy so we can see them? Or maybe even spread them out across the screen so the attack seems larger and more intimidating?

Of course, that's what we do with the pan pot. It gives the engineer (the director) the ability to move the background vocals (Indians) out of the way of the lead vocal (cowboy) so that in this case we can hear (see) each of them much more distinctly.

## PHANTOM CENTER

Stereo, invented in 1931 by Alan Blumlien at EMI Records (the patent wasn't renewed in 1959 when the format was taking off — doh!), features a phenomena known as the *phantom center*. The phantom center means that the output of the two speakers combine to give the impression of a third speaker in between them. This phantom center can sometimes shift as the balance of the music shifts from side to side, which can be very disconcerting to the listener. As a result, film sound has always relied upon a third speaker channel in the center in order to keep the sound anchored. This third channel never caught on in music circles (until now that is. See Part II — Mixing in Surround), mostly because consumers had a hard time finding a place for two speakers, let alone three.

## THE BIG THREE

There are three panoramic areas in the mix that seem to get the most action.

### The Center and the Extreme Hard Left and Right

The center is obvious in that the most prominent music element (usually the lead vocal) is panned there, as well as the kick drum, bass guitar and even the snare drum. Although putting the bass and kick up the middle makes for a musically coherent and generally accepted technique, its origins are really from the era of vinyl records.

When stereo first came into widespread use in the mid-60's, it was not uncommon for mixers to pan most of the music from the band to one side while the vocals were panned opposite. This was because stereo was so new that the recording and mixing techniques for the format hadn't been discovered or refined yet, so pan pots were not yet available on mixing consoles. Instead, a three-way switch was used to assign the track to the left output, right output or both (the center).

Because music elements tended to be hard-panned to one side, this caused some serious problems: if any low frequency boost was added to the music on just that one side, the imbalance in low frequency energy would cause the cutting stylus to cut right through the groove wall when the master lacquer disc (the master record) was cut. The only way around this was to either decrease the amount of low frequency energy from the music to balance the sides, or pan the bass and kick and any other instrument with a lot of low frequency component to the center. In fact, a special equalizer called an Elliptical EQ was used during disc cutting, specifically to move all the low frequency energy to the center in the event that anything with a lot of low frequencies was panned off-center.

Likewise, as a result of the vast array of stereo and pseudo-stereo sources and effects that came onto the market over the years, mixers began to pan these sources hard left and right as a matter of course. Since the mixer's main task is to make things sound bigger and wider, it was an easy choice to pan one of these stereo sources or effects hard left and hard right. Suddenly things sounded huge! The problem came later when almost all keyboards and effects devices came with stereo outputs (many are actually pseudo-stereo with one side just chorused a little sharp then flat against the dry signal). Now the temptation was to pan all of these "stereo" sources hard left and right on top of one another. The result was "Big Mono."

**David Pensado:** *I think that there are three sacred territories in a mix that if you put something there, you've got to have an incredibly good reason. That's extreme left, center and extreme right. I've noticed that some mixers will get stereo tracks from synthesizers and effects and they just instinctively pan them hard left and hard right. What they end up with is these big train wrecks out on the ends of the stereo spectrum. Then they pan their kick, snare, bass and vocals center and you've got all this stuff stacked on top of each other. If it were a visual, you wouldn't be able to see the things behind the things in front. So what I do is take a stereo synthesizer track and I'll just toss one side because I don't need it. I'll create my own stereo by either adding a delay or a chorus or a pre-delayed reverb or something like that to give it a stereo image. I'll pan maybe the dry signal to 10:00 and then I'll pan the effects just inside the extreme left side. I would never put it hard left because then there's too many things on top of it. I would pan it at 9:00, and then pan the dry signal to say 10:30, something like that.*

Big Mono occurs when you have a track with a lot of pseudo-stereo sources that are all panned hard right and hard left. In this case, you're not creating much of a panorama because everything is placed hard left and right and you're robbing the track of definition and depth because all of these tracks are panned on top of one another.

The solution here is to throw away one of the stereo tracks (the chorused one; keep the dry one) and make your own custom stereo patch either with a pitch shifter or delay (see Element Four — Dimension). Then, instead of panning hard left and right, find a place somewhere inside those extremes.

One possibility is to pan the left source to about 10:00 while the right is panned to about 4:00. Another more localized possibility would be to put the left to 9:00 and the right all the way to 10:30. This gives the feeling of localization without getting too wide.

**Ed Seay:** *One of the things I don't like is what I call "big mono," where there's no difference in the left and the right other than a little warble. If you pan that left and right wide, and then here comes another keyboard and you pan that left and right wide and then there's the two guitars and you pan them left and right wide, by the time you get all this stuff left and right wide, there's really no stereo in the sound. It's like having a big mono record and it's just not really aurally gratifying. So to me, it's better to have some segregation and that's one of the ways I try to make everything heard in the mixes. Give everybody a place on the stage.*

## PANNING OUTSIDE THE SPEAKERS

Some mixers like to use the phantom images afforded by some external processors like exciters to pan an instrument outside the speakers. In this case, the phase differences make the instrument seem to come from outside the speakers instead of from inside them. While some find this effect disconcerting, it can be very effective under the right circumstances.

### Panning in Dance Music

**David Sussman:** *If I'm doing a dance club record, I don't go extremely wide with what I consider important elements, which would be kick, snares, hi-hats and cymbals. Because of the venues where the song is being played, if you pan a pretty important element on the left side, half the dance floor's not hearing it. So important elements like that I usually keep either up the middle or maybe like at 10:30 and 1:30. Lead vocals are almost always up the middle.*

### Panning in Mono (Yes, That's Right!)

**Don Smith:** *I check my panning in mono with one speaker, believe it or not. When you pan around in mono, all of a sudden you'll find that it's coming through now and you've found the space for it. If I want to find a place for the hi-hat for instance, sometimes I'll go to mono and pan it around and you'll find that it's really present all of a sudden, and that's the spot. When you start to pan around on all your drum mics in mono, you'll hear all the phase come together. When you go to stereo, it makes things a lot better.*

### Panning for Clarity

**Joe Chiccarelli:** *Once I have my sounds and everything is sitting pretty well, I'll move the pans around a tiny bit. If I have something panned at 3:00 and it's sitting pretty well, I'll inch it a tiny sliver from where I had it just because I found it can make things clearer that way. When you start moving panning around it's almost like EQing something because of the way that it conflicts with other instruments. I find that if I nudge it, it might get out of the way of something or even glue it together.*

# Element Three: Frequency Range — Equalizing

E ven though an engineer has every intention of making his tracks sound as big and as clear as possible during tracking and overdubs, it often happens that the frequency range of some or all of the tracks is still somewhat limited when it comes time to mix. This can be because the tracks were recorded in a different studio using different monitors, a different signal path, or highly influenced by the producer and musicians. As a result, the mixing engineer then must extend the frequency range of those tracks.

In the quest to make things sound bigger, fatter, brighter and clearer, the equalizer is the chief tool used by most mixers. But, perhaps more than any other audio tool, the use of the equalizer requires a skill that separates the average engineer from the master.

**Allen Sides:** *What I would say is that I tend to like things to sound sort of natural but I don't care what it takes to make it sound like that. Some people get a very pre-conceived set of notions that you can't do this or you can't do that. Like Bruce Swedien said to me, he doesn't care if you have to turn the knob around backwards; if it sounds good, it is good. Assuming that you have a reference point that you can trust, of course.*

## WHAT ARE YOU TRYING TO DO?

There are three primary goals when equalizing:

1) To make an instrument sound clearer and more defined

2) To make the instrument or mix bigger and larger than life

3) To make all the elements of a mix fit together better by juggling frequencies so that each instrument has its own predominant frequency range

Before we examine some methods of equalizing, it's important to note the areas of the audio band and what effect they have on what we hear. The audio band can effectively be broken down into six distinct ranges, each one having enormous impact on the total sound.

- **Sub-Bass** — The very low bass between 16Hz and 60Hz that encompasses sounds that are often felt more than heard, such as thunder in the distance. These frequencies give the music a sense of power even if they occur infrequently. Too much emphasis on this range makes the music sound muddy.

- **Bass** — The bass between 60Hz and 250Hz contains the fundamental notes of the rhythm section, so EQing this range can change the musical balance, making it fat or thin. Too much boost in this range can make the music sound boomy.

- **Low Mids** — The midrange between 250Hz and 2000Hz contains the low order harmonics of most musical instruments and can introduce a telephone-like quality to the music if boosted too much. Boosting the 500Hz to 1000Hz octave makes the instruments sound horn-like, while boosting the 1kHz to 2kHz octave makes them sound tinny. Excess output in this range can cause listening fatigue.

- **High Mids** — The upper midrange between 2kHz and 4kHz can mask the important speech recognition sounds if boosted, introducing a lisping quality into a voice and making sounds formed with the lips such as "m," "b" and "v" indistinguishable. Too much boost in this range — especially at 3kHz — can also cause listening fatigue. Dipping the 3kHz range on instrument backgrounds and slightly peaking 3kHz on vocals can make the vocals audible without having to decrease the instrumental level in mixes where the voice would otherwise seem buried.

- **Presence** — The presence range between 4kHz and 6kHz is responsible for the clarity and definition of voices and instruments. Boosting this range can make the music seem closer to the listener. Reducing the 5kHz content of a mix makes the sound more distant and transparent.

• **Brilliance** — The 6kHz to 16kHz range controls the brilliance and clarity of sounds. Too much emphasis in this range, however, can produce sibilance on the vocals.

Leo di Gar Kulka — "Equalization - The Highest, Most Sustained Expression of the Recordist's Heart," *Recording Engineer/Producer*, Vol. 3, Number 6, November/December, 1972

**Figure 3**

| RANGE | DESCRIPTION | EFFECT |
|---|---|---|
| 16 – 60Hz **Sub-Bass** | Sense of power; felt more than heard | Too much makes the music sound muddy |
| 60 – 250Hz **Bass** | Contains fundamental notes of rhythm section; makes music fat or thin | Too much makes the music boomy |
| 250 – 2kHz **Low Mids** | Contains the low order harmonics of most instruments | Boosting 500 – 1kHz sounds hornlike; 1 – 2kHz sounds tinny |
| 2kHz – 4kHz **High Mids** | Contains speech recognition sounds like "m," "b" and "v" | Too much causes listener fatigue |
| 4kHz – 6kHz **Presence** | Responsible for clarity and definition of voices and instruments | Boosting makes music seem closer |
| 6kHz – 16kHz **Brilliance** | Controls brilliance and clarity | Too much causes vocal sibilance |

For those of you who have an easier time visualizing the audio spectrum in one-octave increments (like those found on a graphic equalizer), here's an octave look at the same chart.

**Figure 4**

| | |
|---|---|
| 31Hz | Rumble, "chest" |
| 63Hz | Bottom |
| 125Hz | Boom, thump, warmth |
| 250Hz | Fullness or mud |
| 500Hz | Honk |
| 1KHz | Whack |
| 2KHz | Crunch |
| 4KHz | Edge |
| 8KHz | Sibilance, definition, "ouch!" |
| 16KHz | Air |

Since each specific song and instrument and player is unique, it's impossible to give anything other than some general guidelines as to equalization methods. Also, different engineers have different ways of arriving at the same end, so if the following doesn't work for you, keep trying. The method doesn't matter, only the end result.

Before these methods are outlined, it's really important that you observe the following:

**LISTEN!** **Open up your ears and listen carefully to all the nuances of the sound. It's all-important.**

**Make sure you're monitoring at a comfortable level, not too loud and not too soft**. If it's too soft, you may be fooled by the non-linearity of the speakers and overcompensate. If it's too loud, certain frequencies may be masked or overemphasized by the non-linearity of the ear itself (thanks to the Fletcher-Munson curves) and again you will overcompensate.

## 1) Equalize to make an instrument sound clearer and more defined

Even some sounds that are recorded well can be lifeless, thanks to certain frequencies being overemphasized or others being severely attenuated. More often than not, the lack of definition of an instrument is because of too much lower midrange in approximately the 400Hz to 800Hz area. This area adds a "boxy" quality to the sound.

A) Set the boost/cut knob to a moderate level of cut (8 or 10dB should work).

B) Sweep through the frequencies until you find the frequency where the sound has the least amount of boxiness and the most definition.

C) Adjust the amount of cut to taste. Be aware that too much cut will cause the sound to be thinner.

D) If required, add some "point" to sound by adding a slight amount (start with only a dB; add more to taste) of upper midrange (1kHz to 4kHz).

E) If required, add some "sparkle" to sound by adding a slight amount of high frequencies (5kHz to 10kHz).

F) If required, add some "air" to sound by adding a slight amount of the brilliance frequencies (10kHz to 15kHz).

**PLEASE NOTE!** Always try attenuating (cutting) the frequency first. This is preferable because all equalizers add phase shift as you boost, which results in an undesirable coloring of sound. Usually, the more EQ you add, the more phase shift is also added and the harder it will be to fit the instrument into the mix. Many engineers are judicious in their use of EQ. That being said, anything goes! If it sounds good, it is good.

### Alternate Method

1) Starting with your EQ flat, remove ALL the bottom end by turning the low frequency control to full cut.

2) Using the rest of your EQ, tune the mid-upper midrange until the sound is thick yet distinct.

3) Round it out with a supporting lower-mid tone to give it some body.

4) Slowly bring up the mud-inducing bottom end enough to move air, but not so much as to make the sound muddy.

5) Add some high frequency for definition.

**Ed Seay:** *I just try to get stuff to sound natural, but at the same time be very vivid. I break it down into roughly three areas: mids and the top and the bottom. Then there's low mids and high mids. Generally, except for a very few instruments or a few microphones, cutting flat doesn't sound good to most people's ears. So I'll say, "Well, if this is a state of the art preamp and a great mic and it doesn't sound that great to me, why?" Well, the mid range is not quite vivid enough. Okay, we'll look at the 3k, 4k range, maybe 2500. Why don't we make it kind of come to life like a shot of cappuccino and open it up a little bit? But then I'm not hearing the air around things, so let's go up to 10k or 15k and just bump it up a little bit and see if we can kind of perk it up. Now all that* *sounds good but our bottom is kind of undefined. We don't have any meat down there. Well, let's sweep through and see what helps the low end." Sometimes, depending on different instruments, a hundred cycles can do wonders for some instruments. Sometimes you need to dip out at 400 cycles because that's the area that sometimes just clouds up and takes the clarity away. But a lot of times, adding a little 400 can fatten things up.*

## 2) Equalize to make the instrument or mix bigger and larger than life

*"Bigness" usually comes from the addition of bass and sub-bass frequencies in the 40Hz to 250Hz range. This will come from a region below 100Hz, a region above 100Hz or both.*

A) Set the boost/cut knob to a moderate level of boost (8 or 10dB should work).

B) Sweep through the frequencies in the bass band until you find the frequency where the sound has the desired amount of fullness.

C) Adjust the amount of boost to taste. Be aware that too much boost will make the sound muddy.

D) Go to the frequency either half or twice the frequency that you used in B and add a moderate amount of that frequency as well. Example: If your frequency in B was 120Hz, go to 60Hz and add a dB or so as well. If your frequency was 50Hz, go to 100Hz and add a bit there.

**PLEASE NOTE:**    *1. It's usually better to add a small amount at two frequencies than a large amount at one.*

*2. Be aware that making an instrument sound great while soloed may make it impossible to fit together with other instruments in the mix.*

> *Rule of Thumb — The fewer instruments in the mix, the bigger each one should be. Conversely, the more instruments in the mix, the smaller each one needs to be in order for all to fit together.*

## 3) Equalize to make all the elements of a mix fit together better by juggling frequencies so that each instrument has its own predominant frequency range

A) Start with the rhythm section (bass and drums). The bass should be clear and distinct when played against the drums, especially the kick and snare.

Each instrument should be heard distinctly. If not do the following:

1) Make sure that no two equalizers are boosting at the same frequency. If so, move one to a frequency a little higher or lower.

2) If an instrument is cut at a certain frequency, boost the frequency of the other instrument at that same frequency. Example: The kick is cut at 500Hz. Boost the bass at 500Hz.

B) Add the next most predominant element, usually the vocal and proceed as above.

C) Add the rest of the elements into the mix one by one. As each instrument is added it should be checked against the previous elements as above.

**REMEMBER**: *1. The idea is to hear each instrument clearly and the best way for that to happen is for each instrument to live in its own frequency band.*

*2. After frequency juggling, an instrument might sound terrible when soloed by itself. That's OK, the goal is that it work in the track.*

**Jon Gass:** *I really start searching out the frequencies that are clashing or rubbing against each other. Then I work back towards the drums. But I really try to keep the whole picture in there most of the time as opposed to really isolating things too much. If there are a couple, two or three instruments that are clashing, that's probably where I get more into the solo if I need to kind of hear the whole natural sound of the instrument. I'll try to go more that way with each instrument unless there's a couple that are really clashing, and then I'll EQ more aggressively. Otherwise, I'm not scared to EQ quite a bit.*

**Ed Seay:** *Frequency juggling is important. You don't EQ everything in the same place. You don't EQ 3k on the vocal and the guitar and the bass and the synth and the piano, because then you have such a buildup there that you have a frequency war going on. So sometimes you can say, "Well, the piano doesn't need 3k, so let's go lower, or let's go higher." Or, "This vocal will pop through if we shine the light not in his nose, but maybe towards his forehead." In so doing, you can make things audible and everybody can get some camera time.*

## EASY-TO-REMEMBER GOLDEN RULES OF EQ

> 1. If it sounds muddy, cut some at 250Hz.
> 2. If it sounds honky, cut some at 500Hz.
> 3. Cut if you're trying to make things sound better.
> 4. Boost if you're trying to make things sound different.
> 5. You can't boost something that's not there in the first place.

**Figure 5**

| INSTRUMENT | MAGIC FREQUENCIES |
|---|---|
| Bass Guitar | Bottom at 50 – 80Hz; attack at 700Hz; snap at 2.5kHz |
| Kick Drum | Bottom at 80 – 100Hz; hollowness at 400Hz; point at 3 – 5kHz |
| Snare | Fatness at 120 – 240Hz; boing at 900Hz; crispness at 5kHz; snap at 10kHz |
| Toms | Fullness at 240 – 500Hz; attack at 5 – 7kHz |
| Floor Tom | Fullness at 80 – 120Hz; attack at 5kHz |
| Hi Hat and Cymbals | Clang at 200Hz; sparkle at 8 to 10kHz |
| Electric Guitar | Fullness at 240 – 500Hz; presence at 1.5 to 2.5kHz; reduce 1kHz for 4x12 cabinet sound |
| Acoustic Guitar | Fullness at 80Hz; body at 240Hz; presence at 2 – 5kHz |
| Organ | Fullness at 80Hz; body at 240Hz; presence at 2 – 5kHz |
| Piano | Fullness at 80Hz; presence at 2.5 – 5kHz; Honkey-tonk at 2.5kHz; |
| Horns | Fullness at 120 – 240Hz; piercing at 5kHz |
| Voice | Fullness at 120; boominess at 240Hz; presence at 5kHz; sibilance at 5kHz; air at 10 – 15kHz |
| Strings | Fullness at 240Hz; scratchiness at 7 – 10kHz |
| Conga | Ring at 200Hz; slap at 5kHz |

- General Tips
  Use a narrow Q (bandwidth) when cutting; use wide Q's when boosting

  If you want something to stick out, roll off the bottom; if you want it to blend in, roll off the top

- For Snare — To find the "point" on the snare, boost the upper midrange starting at about +5 or 6dB at 2kHz or so. Open up the bandwidth (if that parameter is available) until you get the snare to jump out, then tighten the bandwidth until you get only the part of the snare sound that you want most. Then fine-tune the frequency until you need the least amount of boost in order to make it jump out of the mix.

- For Drums

**Dave Pensado:** *A lot of the music I do has samples in it and that gives the producer the luxury of pretty much getting the sound he wanted from the start. In the old days you always pulled out a little 400 on the kick drum. You always added a little 3 and 6 to the toms. That just doesn't happen as much any more because when I get the tape, even with live bands, the producer's already triggered the sound he wanted off the live performance and the drums are closer.*

- For Bass — The ratio between the low bass (80–120Hz) and the mid-bass (130Hz–200Hz) is important. Try using two fairly narrow peaking bands, one at 100Hz and another at 140Hz and boost one and cut the other. If the bass is too warm, sometimes reducing the upper band can make it more distinct without removing the deeper fundamentals that live in the 100Hz band. Also, try boosting some of the 1kHz area since this is where a lot of the sound of the Fender bass lives.

- For Fatter Guitars — Boost midrange a lot (9dB or so) and sweep the frequencies until you hear the range where the guitar sounds thick but yet still bright enough to cut through. Now, back the boost down to about +4 or so until the guitar cuts through the mix without being too bright.

**Don Smith:** *I use EQ different from some people. I don't just use it to brighten or fatten something up; I use it to make an instrument feel better. Like on a guitar, making sure that all the strings on a guitar can be heard. Instead of just brightening up the high strings and adding mud to the low strings, I may look for a certain chord to hear more of the A string.*

*If the D string is missing in a chord, I like to EQ and boost it way up to +8 or +10 and then just dial through the different frequencies until I hear what they're doing to the guitar. So I'm trying to make things more balanced in the way they lay with other instruments.*

• For Vocals

Boost a little at 125Hz to 250Hz to accentuate the voice fundamental and make it more "chesty"-sounding. The 2kHz to 4kHz range accentuates the consonants and makes the vocal seem closer to the listener.

**Ed Seay:** *On a vocal sometimes I think, "Does this vocal need a diet plan? Does he need to lose some flab down there?" Or sometimes, "We need some weight on this guy so let's add some 300 cycles and make him sound a little more important."*

**David Sussman:** *If I'm recording vocals, I like to roll off quite a bit on the bottom end so the compressor doesn't start kicking in and bringing up any low end rumble or noise. If I'm EQing a piano or something that's already been recorded, I sometimes roll off a lot of the bottom so I leave a lot of room for the bass and the kick drum to occupy. A lot of times I don't need anything under probably 100Hz. I'll do some rolling off with the filters and then I may take a bell curve and zone in on a couple of other woofy areas on certain instruments.*

**Dave Pensado:** *I think of EQ as an effect much the same way you would add chorus or reverb to a particular instrument or vocal. Like, I might have a vocal where I think it's really EQed nicely and then I'll add a little more 3k just to get it to bite a little more. Then it just makes me feel like the singer was trying harder and it brings out a little bit of passion in his or her voice. So I tend to be most effective when I do the standard equalizing, then take it to the next level, thinking of it as an effect.*

# Element Four: Dimension — Adding Effects

T he fourth element of a mix is *dimension,* which is the ambient field in which the track or tracks sit. Dimension can be captured while recording but usually has to be created or enhanced when mixing by adding effects such as reverb or delay or any of the modulated delays such as chorusing or flanging. Dimension might be something as simple as recreating an acoustic environment, but it also could be the process of adding width or depth to a track or trying to spruce up a boring sound.

Actually, there are really four reasons why a mixer would add dimension to a track:

- **To create an aural space**

- **To add excitement**

**Joe Chiccarelli:** *I try to start out with a flat track, then find the tracks that are boring and add some personality to them.*

- **To make a track sound bigger, wider and/or deeper**

**Lee DeCarlo:** *Everything has to be bigger always. Now, a lot of times I'll do stuff with no effect on it whatsoever, but I don't particularly like it. Effects are makeup. It's cosmetic surgery. I can take a very great song by a very great band and mix it with no effects on it at all and it'll sound good, and I can take the same song and mix it with effects and it'll sound fucking fantastic! That's what effects are for. It's just makeup.*

- **To move a track back in the mix (give the impression it's farther away)**

**Dave Pensado:** *The way I think of it is the pan knob places you left to right while the effects tend to you place you front to rear. That's a general statement, but it's a good starting point. In other words, if you want the singer to sound like she's standing behind the snare drum, leave the snare drum dry and wet down the singer and it'll sound like the singer is standing*

*Chapter Six*

*that far behind the snare drum. If you want the singer in front of the snare drum, leave him dry and wet down the snare drum.*

One of the reasons why we record elements in stereo is to capture the natural ambience (or dimension) of an instrument. Since we can't record everything this way due to track or storage limitations, we must create this aural space artificially.

**Ed Seay:** *Sometimes [I add effects for] depth and sometimes you just want it to sound a little bit more glamorous. Other times you just want it to sound appropriate. Well, appropriate to what? If it's an arena rock band, then all this room stuff is going to make it sound like they flunked out of the arena circuit and they're now doing small clubs. But if you got a band where that's more of an in-your-face, hard-driving thing, you want to hear the room sound.*

*I've done records where I didn't use any effects or any verb, but quite often just a little can make a difference. You don't even have to hear it but you can sense it when it goes away. It's just not quite as friendly sounding, not quite as warm. Obviously an effect is an ear catcher or something that can just kind of slap somebody and wake them up a little bit in case they're dozing off there.*

Although there are no specific rules for these, there are some guidelines.

• **As a general rule of thumb, try to picture the performer in an acoustic space and then realistically recreate that space around them.**
This method usually saves some time over simply experimenting with different effects presets until something excites you (although if that method works for you, that's what you should continue to do). Also, the created acoustic space needn't be a natural one. In fact, as long as it fits the music, the more creative the better.

• **Smaller reverbs or short delays make things sound bigger.**
Reverbs with decays under a second (usually much shorter than that) and delays under 100 milliseconds (again, usually a lot shorter than that) tend to create an acoustic space around a sound, especially if the reverb or delay is stereo.

Many times a reverb will be used with the decay turned down as far as it will go and this setting is sometimes the most difficult for a digital reverb unit to reproduce, resulting in a metallic sound. If this occurs, sometimes lengthening the decay time a little or trying a different preset will result in a smoother, less tinny sound, or try another unit that performs better under these conditions.

## EQING REVERBS

From the early days of reverb chambers and plates, it's always been common to EQ the reverb returns, although the reasons for doing this have changed over the years. Back when plates and chambers were all that was available, usually some high-frequency EQ at 10kHz or 15kHz was added because the plates and chambers tended to be dark sounding and the reverb would get lost in the mix without the extra high frequency energy.

Nowadays, EQ is added to reverb in order to help create some sonic layering. Here are some points to consider when EQing a reverb return. The type of reverb (digital, real plate, etc.) doesn't matter as much as how these are applied, and that depends on your ears and the song.

Figure 6

### Equalization Tips for Reverbs and Delays

- To make an effect stick out, brighten it up.

- To make an effect blend in, darken it up (filter out the highs).

- If the part is busy (like with drums), roll off the low end of the effect to make it fit.

- If the part is open, add low end to the effect to fill in the space.

- If the source part is mono and panned hard to one side, make one side of the stereo effect brighter and the other darker. (Eddie Van Halen's guitar on the first two Van Halen albums comes to mind here.)

## SONIC LAYERING OF EFFECTS

Sonic layering means that each instrument or element sits in its own ambient environment and this environment is generally created artificially by effects. The idea here is that these sonic atmospheres don't clash with one another, just like in the case of frequency ranges.

Figure 7 features some suggestions to make sure the sonic environments don't clash.

Figure 7

> ## Layering Tips for Reverbs and Delays
>
> * Layer reverbs by frequency with the longest being the brightest and the shortest being the darkest.
>
> * Pan the reverbs any way other than hard left or right.
>
> * Return the reverb in mono and pan accordingly. All reverbs needn't be returned in stereo.
>
> * Get the bigness from reverbs and depth from delays, or vice versa.
>
> * Use a bit of the longest reverb on all the major elements of the track to tie all the environments together.

* **Long delays, reverb pre-delays or reverb decay push a sound farther away if the level of the effect is loud enough.**
  As stated before, delays and pre-delays longer than 100ms (although 250 is where it really kicks in) are distinctly heard and begin to push the sound away from the listener. The trick between something sounding big or just distant is the level of the effect. When the decay or delay is short and the level loud, the track just sounds big. When the decay or delay is long and the level loud, the track just sounds far away.

**Jon Gass:** *I hardly ever use long halls or long reverbs. I use a lot of gear but it's usually for tight spaces. Sometimes in the mix it doesn't sound like I'm using anything, but I might use 20 different reverb type boxes, maybe not set on reverbs, just to create more spaces. I think that helps with the layering and adding textures to things. Though you may not hear it in the mix, you can feel it.*

* **If delays are timed to the tempo of the track, they add depth without being noticeable.**
  Most engineers set the delay time to the tempo of the track (Figure 8 shows how to do this). This makes the delay pulse with the music and adds a reverb type of environment to the sound. It also makes the delay seem to disappear as a discrete repeat but still adds a smoothing quality to the sound.

**Don Smith:** *I usually start with the delays in time, whether it's eighth note or quarter note or dotted value or whatever. Sometimes on the drums I'll use delays very subtly. If you can hear them, then they're too loud; but if you turn them off, you definitely know they're gone. It adds a natural slap — like in a room, so to speak — that maybe you won't hear but you feel. And, if the drums are dragging, you can speed the delays up just a nat so the drums feel like they're getting a lift. If they're rushing, you can do it the other way by slowing the delays so it feels like they're pulling the track back a bit.*

Delays are measured tempo-wise using musical notes in relation to the tempo of the track. In other words, if the song has a tempo of 120 beats per minute (bpm), then the length of time it takes a quarter note to play would be one-half second (60 seconds/120bpm = .5 seconds). Therefore a quarter note delay should be .5 seconds or 500ms (.5 X 1000), which is how almost all delay devices are calibrated.

But 500ms might be too long and set the source track too far back in the mix. Divide that in half for an eighth note delay (500ms/2 = 250ms). Divide in half again for a sixteenth note delay (250ms/2 = 125ms). Divide again for a 1/32nd note delay (125/2 = 62.5ms or rounded up to 63). That still might not be short enough for you so divide again for 1/64th note (62.5/2 = 31.25 or rounded to 31ms). Again, this might not be short enough so divide again for a 1/128th note (31ms/2 = 15.625 rounded up to 16ms). This still might not be short enough, so divide again for a 1/256th note, if there is such a thing (16ms/2 = 8ms).

Now such small increments like 8ms and 16ms might not seem like much, but they're used all the time to make a sound bigger and wider. Even a short delay like this will fit much more smoothly into the track if it's timed.

**Bruce Swedien:** *I think that's (early reflections of a sound) a much overlooked part of sound because there are no reverb devices that can generate that. It's very important. Early reflections will usually occur under 40 milliseconds. It's a fascinating part of sound.*

It's also possible (and sometimes even preferable) to use other note denominations such as triplets or dotted eighths, sixteenths, etc. These can be figured out using the following formula:

**Delay Time x 1.5 = Dotted Value**
Example: 500ms (quarter note 120bpm delay) x 1.5 = 750ms
(Dotted Quarter Note)

**Delay Time x .667 = Triplet Value**
Example: 500ms (quarter note 120bpm delay) x .667 = 333.5ms
(Quarter Note Triplet)

As with the straight notes (quarter, eighths, etc.), you can continually divide the above values in half until you get the desired denomination.

## CALCULATING THE DELAY TIME

Once the beats per minute rate (see Figure 8) is known, most engineers determine the delay time by looking at a chart that identifies the delay time at any rate (there's a chart in the back of this book), by using a utility program found on some computers (*StudioCalc* on the Mac is a popular one), or by using a tap function found on many effects devices (like the Lexicon PCM 80 or 90). When none of these are available, you can still determine the delay time by using a little math.

**60,000/Song Tempo in bpm = Quarter Note Delay in Milliseconds.**

All the other values can be determined from this by doing any of the following:

• Dividing by 2 for lower denominations

• Multiplying any of the above by 1.5 for dotted values

• Multiplying any of the above by 667 for triplet values

**Figure 8**

## Calculating the Delay Time

1) Start a stopwatch when the song is playing and count 25 beats

2) Stop the stopwatch on the 25th beat and multiply the time by 41.81

*The result is the delay time in milliseconds for a quarter note delay.*

or

$$\frac{60,000}{\text{Song Tempo (in beats per minute)}}$$

*The result equals the delay time in milliseconds for a quarter note delay.*

- **If delays are not timed to the tempo of the track, they stick out.** Sometimes you want to hear a delay distinctly and the best way to do that is to make sure that the delay is NOT exactly timed to the track. Start by first putting the delay in time with the track, then slowly alter the timing until the desired effect is achieved.

- **Reverbs work better when timed to the tempo of the track.** Reverbs are timed to the track by triggering them off of a snare hit and adjusting the decay parameter so that the decay just dies by the next snare hit. The idea is to make the decay "breathe" with the track.

The best way to achieve this is to make everything as big as possible at the shortest setting first, then get gradually longer until it's in time with the track.

The pre-delay of a reverb (the space between where the note of the source track dies off and the reverb begins) can change the sound of the reverb considerably and is usually timed to the tempo of the track. Back in the days of real plates and chambers, pre-delay was achieved by using a tape slap (see Figure 9). This was the natural echo that occurred when playing back off the repro head of a tape machine while recording onto it. Since there was a gap between the record and playback head, it gave a noticeable delay, and early engineers used this to their advantage. Because the early tape machines didn't have varispeed, it wasn't possible to time the delay to the tempo of the track. The best that could be done was to select either a 7 1/2 ips or 15 ips tape slap.

**Figure 9  Pre-delay Using Tape Slap**

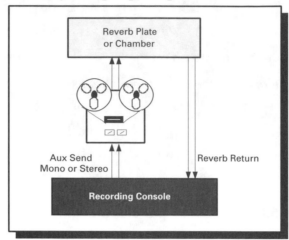

**Tape-Based Pre-delay**

7 1/2 ips = 250ms
15 ips = 125ms
Times are approximate since the gap between the record and
playback head is slightly different on each model of tape
machine.

## RE-AMPING

One of the ways that a natural environment is recreated is a
process known as *re-amping*. This is accomplished by actually
sending a signal of an already recorded track (say a guitar)
back out to an amplifier in the studio and re-miking it from a
distance in order to capture the ambience of the room. It's
all the better if the ambience is recorded in stereo.

**Bruce Swedien:** *What I will do frequently when we're layering with synths and so on is*
*to add some acoustics to the synth sounds. I think this helps in the*
*layering in that the virtual direct sound of most synthesizers is not too*
*interesting. So I'll send the sound out to the studio and use a coincident*
*pair of mics to blend a little bit of acoustics back with the direct sound.*
*Of course, it adds early reflections to the sound, which reverb devices*
*can't do. That's the space before the onset of reverb where those early*
*reflections occur.*

Even though a multitude of digital effects boxes on the market have a flanging preset, almost nothing sounds like the real thing. Flanging (another name for an artificially induced comb filter) got its name from the fact that the effect is achieved by actually slowing down a reel of tape by holding your finger on the edge of the reel flange (the metal piece on each side of the tape that holds the reel together). The effect was first noticed by the public on The Small Faces 1966 hit "Itchycoo Park" (it's actually been reported to have been invented by Les Paul in the 50's) but used extensively by The Beatles, Jimi Hendrix and many others of that time.

**The Vintage Method (see Figure 10)**

1) Play the recording to be processed on Deck 1.
2) Split Deck 1's output to Deck 2 and Deck 3.
3) Mix Deck 2 and Deck 3's outputs together and record on Deck 4 (the master).
4) Set Decks 1, 2, and 3 in Repro Monitor mode. Set Deck 4 in Input Monitor mode.
5) Start Decks 2, 3, and 4 in Record. Start Deck 1 in Play.
6) Put your finger on a flange of a supply reel on Deck 2 or 3 and flange away.
7) Splice the Flanged Master into the original master

The flanged master is two generations away from the original master. The first generation is in the recording to Decks 2 and 3; the second generation is the recording that saves your flanging work on Deck 4.

Remember that you'll need to drop the output level of the two machines 1.5dB each, since you will add 3dB to the final level when you combine the signals of the two decks. Otherwise, the level will jump when you cut the flanged portion back into the original master.

**Figure 10  Tape Flanging**

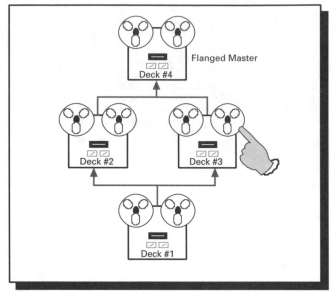

## The Modern Method (see Figure 11)

This is almost the same except that Deck 1 is a multitrack sending an automated mix to Deck 2.
If you know the delay time of the headblock of Deck 2, you can substitute a good phase-locked stereo delay for Deck 3.
Deck 4 can be a DAT deck or can even be eliminated if the flanged result is recorded back onto the multitrack.

**Figure 11  Modern Tape Flanging**

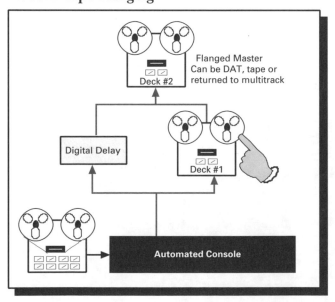

- **For Fatter Lead Background Vocals** — Use some chorusing (very short modulated delays) panned hard left and right to fatten up the sound. Use different EQ and reverb settings on the delays. (Make sure you check the mix in mono to be sure that the delays aren't canceling.) Ride the chorusing effect, adding and subtracting it according to what sounds best.

- **For Out of Tune Vocals** — Use a stereo pitch shifter with one side tuned slightly high and the other tuned slightly low. Pan these left and right. The more out of tune the vocal, the more you might want to de-tune the pitch up and down. This does an effective job of taking the listener's attention off the sour notes.

- **For Electronic Keyboards** — A nice delay effect that simulates a small room can be achieved by using a stereo delay and setting the delay times to 211ms and 222ms.

- **For Fatter Guitars** — Delay the guitar about 12ms (or whatever the tempo dictates) and hard pan both the guitar and delay. This sounds like two people playing perfectly in sync, yet sounds bigger and still keeps a nice hole open in the middle for the vocals.

- **For Fatter Guitars (2)** — Pan the guitar track and the delay to the center (or put your monitors in mono), then slowly increase the delay time until it sounds bigger. Increase it a little more for good measure. You'll probably find the result is 25–30ms.

- **For Fatter Guitars (3)** — For years, LA session guitarists have automatically dialed up a stereo delay of 25ms on one side and 50ms on the other.

**Allen Sides:** *I'm a big fan of the RMX16, not for drums, but for vocals and guitars and stuff. I love non-lin for guitars and things. Let's say that you had a couple of discrete guitars that were playing different lines and you try putting them in the middle and they get on top of each other. If you put them left and right, they're too discrete. The RMX non-lin set at 4 seconds with a 10 millisecond pre-delay and an API EQ on the send with about +4 at the 12k shelf and -2 at 100 Hz going into it does a wonderful job of creating a left/right effect, but it still spreads nicely. It works great for that.*

- **For Tommy Lee "Thunder Drums"** — For this to work, the bass drum has to sound tight to begin with, with a decent amount of beater present, and all the drums should be gated with the gate timed to the track. Set a reverb on the "cathedral" or "large hall" setting and then add a little to all parts of the kit. Pan the reverb returns to sit the reverb sound behind each part of the kit.

# CHAPTER 7

# Element 5: Dynamics — Compression and Gating

I n years past, the control of the volume envelope of a sound (*dynamics*) would not have been included as a necessary element of a great mix. In fact, dynamics control is still not a major part of Classical and Jazz mixing. But in today's modern music, the manipulation of dynamics plays a major role in the sound. In fact, just about nothing else can affect your mix as much and in so many ways as compression.

**Jerry Finn:** *I think that the sound of modern records today is compression. Audio purists talk about how crunchy compression and EQ is, but if you listen to one of those Jazz or Blues records that are done by the audiophile labels, there's no way they could ever compete on modern radio even though they sound amazing. And unfortunately, all the phase shift and pumping and brightening and stuff that's imparted by EQ and compression is what modern records sound like. Every time I try to be a purist and go, "You know, I'm not gonna compress that," the band comes in and goes, "Why isn't that compressed?"*

## DYNAMICS CONTROLLERS

Dynamics are controlled by the use of compression, limiting and gating. For those of you new to mixing or need a review or clarification, here's a brief description of each. See the glossary or any number of recording texts for more complete information.

## COMPRESSION

Compression is an automated level control, using the input signal itself to determine the output level. This is set by using the **threshold** and **ratio** controls. Compressors work on the principle of gain ratio, which is measured on the basis of input level to output level (see Figure 12). For example, this means that for every 4dB that goes into the compressor, 1dB will come out for a ratio of four to one (normally written as 4:1). If a gain ratio of 8:1 was set, then for every 8dB that goes into the unit, only 1dB will come out. Although this could apply to the entire signal regardless of level, a compressor is usually not set up that way. A **threshold** control determines at what signal level the compressor will begin to operate. Therefore, threshold and ratio are inter-related and one will affect the way the other works. Some compressors (like LA2-As and UREI LA-3s) have a fixed ratio, but on most units the control is variable.

Most compressors also have **attack** and **release** parameters. These controls determine how fast or slow the compressor reacts to the beginning (attack) and end (release) of the signal. Many compressors have an Auto mode that sets the attack and release in relation to the dynamics of the signal. Although Auto gener-ally works relatively well, it still doesn't allow for the precise settings required by certain source material. Some compressors (like the dbx 160 series) have a fixed attack and release which gives it a particular sound.

When a compressor operates it actually decreases the gain of the signal so there is another control called **make-up gain** or **output**, which allows the signal to be boosted back up to its original level or beyond.

**Figure 12**

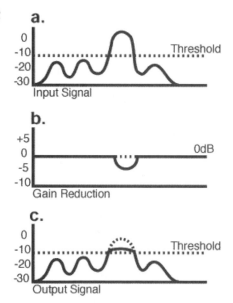

Illustration courtesy of Alesis Studio Electronics

*The Mixing Engineer's Handbook*

In Figure 12, Diagram A shows an input signal before compression.
Diagram B shows what the compressor does — turning down the volume once your signal crosses the threshold.
Diagram C shows the new output level (with the original signal shown as a dotted line).

## LIMITING

The same box can do compression and limiting for the most part. The difference is how they're set up. Any time the compression ratio is set to 10:1 or more, the result is considered limiting. A limiter is essentially a brick wall for level, allowing the signal to get only to a certain point and little more. Think of it as the same thing as a governor that's sometimes used on trucks to make sure that they don't go over the speed limit. Once you hit 55mph (or whatever the speed limit is in your state), no matter how much you depress the gas pedal, you won't go any faster. Same with a limiter. Once you hit the predetermined level, no matter how much you try to go beyond it, the level pretty much stays the same.

Limiting is usually used in sound reinforcement for speaker protection (there are some limiters on powered studio monitors as well), and not used much in mixing with the following exception:

*Many engineers who feel that the bass guitar is the anchor for the song want the bass to have as little dynamic range as possible. In this case, limiting the bass by 3–6dB (depending on the song) with a ratio of 10:1, 20:1 or even higher will achieve that.*

## GATING

Although not used nearly as much now that console automation is so prevalent, gates are still a major player in the mixer's arsenal. A gate keeps a signal turned off until it reaches a threshold level, then the gate opens and lets the sound through. The gate can be set to turn the sound completely off when it drops below threshold or lower the level a predetermined amount. Depending on the situation, just turning the level down a bit sounds more natural than turning it completely off, although completely off can be used as a great effect.

A gate (sometimes called *noise gate* or *expander*) is usually used to cover up some problems on a track such as noises, buzzes, coughs or other low-level noises off mic. On loud guitar tracks for instance, a gate can be used to effectively get rid of amplifier noise when the guitar player is not playing. On drums, gates can be used to turn off the leakage from the tom mics, which tend to muddy up the other drum tracks. A gate can also be used to tighten up the sound of a floppy kick drum by decreasing the after-ring.

## WHY ADD COMPRESSION?

If there is one major difference between the sound of a demo or semi-pro recording and a finished professional mix, it's the use of compression. As a matter of fact, the difference between one engineer's sound and another's is, more often than not, in his use of compression.

**George Massenburg:** *The big difference between engineers today is the use of the compressor. At one time or another I tried to compress everything because I was building a compressor and I wanted to see how it did on every instrument. I'm a little off on compression now because there are so many people that over-use it. Everything is squeezed to death.*

There are two reasons to add compression to a track or mix: to control the dynamics or as an effect.

## CONTROLLING DYNAMICS

Controlling dynamics means keeping the level of the sound even. In other words, lifting the level of the soft passages and lowering the level of the loud ones so that there is not so much of a difference between them.

Here are a couple of instances where this would be useful:

• **On a bass guitar** – Most basses inherently have certain notes that are louder than others and some that are softer than others. Compression evens out these differences.

• **On a lead vocal** – Most singers can't sing every word or line at the same level so some words may get buried as a result. Compression allows every word to be heard.

- **On a kick or snare drum** – Sometimes the drummer won't hit every beat with the same intensity. Compression can make all drum hits sound relatively the same.

When controlling dynamics, usually a very small amount of compression (2dB to 4dB or so at a 2:1 to 4:1 ratio) is used to limit the peaks of the signal.

**Benny Faccone:** *I like to compress everything just to keep it smooth and controlled, not to get rid of the dynamics. Usually I use around a 4:1 ratio on pretty much everything I do. Sometimes on guitars I go to 8:1. On the kick and the snare I try not to hit it too hard because the snare really darkens up. It's more for control, to keep it consistent. On the bass, I hit that a little harder, just to push it up front a little more. Everything else is for control more than sticking it right up in your face.*

**David Pensado:** *I very rarely use a compressor to even out dynamics. Dynamics are something that I just can't get enough of. The compressors I like the most tend to be the ones that actually help me get dynamics. That might be a contradictory statement, but if you're careful with the attack and release times, you can actually get a compressor to help you with it.*

## COMPRESSION AS AN EFFECT

Compression can also radically change the sound of a track. A track compressed with the right unit and with the correct settings can make the track seem closer and have more aggression and excitement. The volume envelope of a sound can be modified to have more or less attack, which will make it sound punchy, or more of a decay so it sounds fatter.

**Andy Johns:** *I use compression because it's the only way that you can truly modify a sound because whatever the most predominant frequency is, the more you compress it the more predominant that frequency will be. Suppose the predominant frequencies are 1 to 3k. Put a compressor on it and the bottom end goes away, the top end disappears and you're left with "Ehhhh" [makes a nasal sound]. So for me, compressors can modify the sound more than anything else can. If it's a bass guitar, you put the compressor before your EQ, because if you do it the other way around, you'll lose the top and mids when the compressor emphasizes the spot that you EQed. If you compress it first, then add bottom, then you're gonna hear it better.*

One of the little tricks that seems to set New York mixers apart from everyone else is something I call the "New York Compression Trick." It seems like every mixer who's ever mixed in New York City comes away with this maneuver. Even if you don't mix in NYC, once you try it you just might find yourself using this trick all the time, since it is indeed a useful method to make a rhythm section rock.

Here's the trick:

1) Buss the drums, and maybe even the bass, to a stereo compressor.

2) Hit the compressor fairly hard, at least 10dB or more if it sounds good.

3) Return the output of the compressor to a pair of fader inputs on the console.

4) Add a pretty good amount of high end (6–10dB at 10kHz or so) and low end (6–10dB at 100Hz or so) to the compressed signal.

5) Now bring the fader levels of the compressor up until it's tucked just under the present rhythm section mix to where you can just hear it.

The rhythm section will now sound bigger and more controlled without sounding overly compressed.

**Joe Chiccarelli:** *What I will do a lot is buss my drums to another stereo compressor, usually a Joe Meek SC2, and blend that in just under the uncompressed signal. Sometimes what I'll do if everything sounds good but the bass and kick drum aren't locked together or big enough to glue the record together, I'll take the kick and bass and buss them together to a separate compressor, squish that a fair amount and blend it back in. I'll add a little bottom end to that if the record still isn't big enough on the bottom. This helps fit the bass and kick lower on the record and gets it out of the way of the vocal.*

## COMPRESSION ON INDIVIDUAL INSTRUMENTS

In these days with consoles that contain compressors on every channel, it's not uncommon for at least a small amount of compression to be used on every instrument (depending on the music, of course). Once again, compression is used on individual instruments either to control the dynamic range or as an effect.

**Ed Seay:** *To me, the key to compression is that it makes the instrument sound like it's being turned up, not being turned down. If you choose the wrong compressor or you use it the wrong way, then your stuff can sound like it's always going away from you. If you use the correct compressor for the job, you can make it sound like, "Man, these guys are coming at you." It's very active and aggressive.*

**John X:** *I use it a lot. Not always in great amounts but I tend to try to get some handle on the peaks. Loops I rarely mess with. If somebody's got a loop and a certain groove that they like, I almost always leave those things alone because they start getting real squirrelly if you mess with them. All of a sudden the groove can change radically. Anything else, I don't mind slammin' the hell out of as long as it sounds the way I want it to sound. I don't even have a rule about it.*

## COMPRESSION ON THE MIX BUSS

Along with compressing individual tracks, many engineers place a stereo compressor across the mix buss to affect the entire mix as well. Originally, this came about when artists began asking why their mixes sounded different in the studio from what they heard on the radio or when their record (it was still vinyl in those days) came from the pressing plant. Indeed, both the radio and record did sound different because an additional round (or two) of compression was added in both mastering and broadcast. In order to simulate what this would sound like, mixing engineers began to add a touch of compression across the mix buss. The problem was, everybody liked it, so now the majority of records now have at least a bit (a few dB) of compression added to the stereo mix despite the fact that it will probably be recompressed again at mastering and yet again when played on radio or television.

**Joe Chiccarelli:** *Compression is like this drug that you can't get enough of. You squish things and it feels great and it sounds exciting but the next day you come back and it's like the morning after and you're saying, "Oh God, it's too much." So I've been trying to really back it off, especially with stereo buss compression.*

In the case of buss compressors, not all are up to the task. Since usually only 2 or 3dB of compression is added, the compressor itself actually adds an intangible sonic quality. Current favorites are the Fairchild 670 (at a hefty $25,000 each), the Manely Vari-mu, CraneSong STC-8 or the Neve 33609.

**Don Smith:** *Generally, the stereo buss itself will go through a Fairchild 670 (serial #7). Sometimes I'll use a Neve 33609 depending on the song. I don't use much; only a dB or two. There's no rule about it. I'll start with it just on with no threshold just to hear it.*

The sound of a great many (some say the majority) of records in the 80's and 90's comes from the sound of the built-in buss compressor on an SSL console. This is an aggressive compressor with a very distinct sonic signature.

**Kevin Killen:** *I tend to be quite modest on compression because my rationale is that you can always add more but you can never take it off. Since it will probably be applied at a later point during mastering and broadcast, I tend to err on the side of caution.*

Since SSL's hit the marketplace, I know what a temptation it is to set up the quad buss compressor even before you start your mix. I tried that for a while but I found out that I didn't like the way it sounded. What I came up with instead was almost like side-chain compression, where you take a couple of groups on the console and you assign various instruments to them and use a couple of compressors across the buss and mix it in, almost as an effect, instead of using compressors across the inserts. You actually get a sense that there is some compression, yet you can ride the compression throughout the song; so if there's a section where you really want to hear it, like in the chorus, you can ride the faders up.

In most modern music, compressors are used to make the sound punchy and in-your-face. The trick to getting the punch out of a compressor is to let the attacks through and play with the release to elongate the sound. Fast attack times are going to reduce the punchiness of a signal, while slow release times are going to make the compressor pump out of time with the music.

Since the timing of the attack and release is so important, here are a few steps to help set it up. Assuming you have some kind of constant meter in the song, you can use the snare drum to set up the attack and release parameters. The method will work the same for other instruments as well.

1) Start with the slowest attack and fastest release settings on the compressor.

2) Turn the attack faster until the instrument (snare) begins to dull. Stop at that point.

3) Adjust the release time so that after the snare hit, the volume is back to 90–100 percent normal by the next snare beat.

4) Add the rest of the mix back in and listen. Make any slight adjustments to the attack and release times as needed.

**The idea is to make the compressor "breathe" in time with the song.**

**Lee DeCarlo:** *I just get the bass and drums so they just start to pump, to where you can actually hear them breathing in and out with the tempo of the song. What I'll do is put the drums and bass in a limiter and just crush the hell out of it. Then I'll play with the release and the attack times until I can actually make that limiter pump in time with the music. So when the drummer hits the snare, it sucks down and you get a good crest on it, and when he lets go of the snare, the ambience of the bass and the drums suck and shoot back up again. You can actually hear a [breathing sound] going on that was never there before. But it was there, it's just that you're augmenting it by using that limiter.*

**Jerry Finn:** *I would say 10 or 12dB and at a ratio anywhere from like 4:1 to 8:1. My compression technique is something I actually learned from Ed Cherney. He was telling me about compressing the stereo buss when I was assisting him, but I use the same technique on everything. I set the attack as slow as possible and the release as fast as possible so all the transients are getting through and the initial punch is still there, but it releases instantly when the signal drops below threshold. I think that's a lot of the sound of my mixes. It keeps things kinda popping the whole time. Also, you can compress things a little bit more and not have it be as audible.*

### Amount of Compression

The amount of compression added is usually to taste. But generally speaking, the more compression the greater the effect. Less compression (6dB or less) is more for controlling dynamics than for the sonic quality. It is not uncommon for radical amounts of compression to be used, however. About 15 or 20dB is routinely used for electric guitars, room mics, drums and even vocals. As with most everything else, it depends on the song.

## TRICKS AND TIPS

- **For Snare** — It's often useful to gate the effects send on the snare so it only triggers with forceful hits. Send the snare direct out of its channel to another channel on the board and gate this new channel. This channel generally is not sent to the main mix, but can be. You can then EQ the new channel and send it to a reverb/EFX unit. By adjusting the threshold you can control the signal sent to the effects unit. This simple technique allows a different effect to be placed on the snare during harder hits and prevents leakage to the effect during things such as tom hits and kick drumbeats.

**Ed Stasium:** *What I do a lot is take a snare drum and go through an LA-2, just totally compress it, and then crank up the output so it's totally distorted and edge it in a little bit behind the actual drum. You don't notice the distortion on the track, but it adds a lot of tone in the snare, especially when it goes [makes an exploding sound]. Actually, something I've done for the last 20 years is to always split the kick drum and snare drum on a mult and take the second output into a Pultec into a dbx 160VU and into a Drawmer 201 gate. Then I pretty much overemphasize the EQ and compression on that track and use it in combination with the original track.*

- **For Drums** — When gating drums, set the range so it attenuates the signal only about 10 or 20dB. That lets some of the natural ambience remain and prevents the drums from sounding choked.

- **For Piano** — If you liked the early Elton John piano sound, put the piano into two LA-2A's or similar compressors and compress the signal a large amount (at least 10dB). Then put the output into two Pultecs or similar equalizers. Push 14kHz all the way up and set 100Hz to about 5. The effect should be a shimmering sound. The chords hold and seem to chorus.

- **For Vocal** — A good starting point for a lead vocal is a 4:1 ratio, medium attack and release, and the threshold set for about 4 to 6dB of gain reduction.

**Don Smith:** *I'll experiment with three or four compressors on a vocal. I've got a mono Fairchild to Neve's to maybe even a dbx 160 with 10dB of compression to make the vocal just punch through the track.*

- **For Bass** — Using a dbx 160X set at a ratio of infinity:1 (the highest ratio), set the threshold for a 3dB or 4dB reduction. This will keep the bass solid and unmoving in the mix.

- **For Guitar** — Higher ratios of 8:1 or 10:1 sometimes work well, with the threshold set so that the guitar cuts through the track. Attack and release should be timed to the pulse of the song.

**Don Smith:** *I may go 20:1 on a (UREI) 1176 with 20dB of compression on a guitar part as an effect. In general, if it's well recorded, I'll do it just lightly for peaks here and there.*

# Element Six: Interest — The Key to Great ( as opposed to merely good ) Mixes

A lthough having control of the previous five elements may be sufficient for many types of audio jobs and might be just fine to get a decent mix, most popular music requires a mix that can take it to another level. Although it's always easier with great tracks, solid arrangements and spectacular playing, a great mix can take simply OK tracks and transform them into hit material so compelling that people can't get enough of it. It's been done on some of your all-time favorite songs.

**Ed Seay:** *The tough part, and the last stage of the mix, is the several hours it takes for me to make it sound emotional and urgent and exciting so that it's just not a song, it's a record. It's not making it just sound good, it's making it sound like an event. Sometimes that means juggling the instruments or the balances or adding some dynamics help. That's the last stage of when I mix, and that's the part that makes it different or special.*

### So How Can We Get to That Point?

More than being just technically correct, a mix must be as interesting as a good movie. It must build to a climax while having points of tension and release to keep the listener subconsciously involved. Just as a film looks bigger than life, a great mix must sound bigger than real life. The passion and the emotion must be on a level where the listener is sucked in and forced to listen.

### Which Brings Us Back to Where We Started

> **Figure out the direction of the song.**
> **Develop the groove and build it like a house.**
> **Find the most important element and emphasize it.**

## THE DIRECTION OF THE SONG

The first thing that the mixer must do before diving headfirst into the mix is to determine the direction of the song. That is determined by both the artist and the performances. For instance, if the song is folksy in nature then it probably won't need big, bombastic drums and long reverbs and delays. But if the artist is a loud arena rock band then you probably won't want a close, intimate sound.

Although it's absolutely possible to change the direction of the song and have a hit (the Dance version of Amy Grant's "Baby, Baby" comes to mind), usually a song will work best with one artist only one way. A good example of this is Marvin Gaye's "Heard It Through the Grapevine," which has been a hit by many artists in innumerable styles. The direction of Creedence Clearwater is very different from the direction of Gladys Knight and the Pips, yet it works equally well for both. The direction is a function of the artist and the performance.

## DEVELOP THE GROOVE AND BUILD IT LIKE A HOUSE

ALL good music, regardless of whether it's Rock, Jazz, Classical, Rap or some new space music that we haven't heard yet, has a strong groove. *The groove is the pulse of the song and how the instruments dynamically breathe with it.*

We usually think of the groove as coming from the rhythm section (especially the drums) but that's not necessarily always the case. In the Police's "Every Breath You Take," the rhythm guitar establishes the groove, while in most songs from Motown's golden age by the Supremes, Temptations and Four Tops, the groove was established by James Jamerson's bass.

The trick for the mixer is to determine which instrument defines the groove; then build the rest of the mix around it.

## FIND THE MOST IMPORTANT ELEMENT AND EMPHASIZE IT

Equally as meaningful, and in some cases even more important than the groove, is finding whatever element is the most important to the song. In some cases (like Dance and Rap music), the most important element is the groove. Yet in other genres (like Country), it is the vocal.

*Chapter Eight*     **59**

Even though the most important element is often the lead vocal, it doesn't necessarily have to be. It could be a riff, like from The Stones' "Satisfaction" and "Start Me Up" or the Rick James' loop from Hammer's "Can't Touch This." It is always a part so compelling that it forces you to listen to the song.

Whatever part is most important, the mixer must identify it and emphasize it in the mix in order for the mix to be elevated to beyond the ordinary.

**Ed Seay:** *I try to find what's important in the mix. I try to find out if the lead vocal is incredibly passionate, then make sure that the spotlight shines on that. Or if the acoustics are sitting there but they're not really driving the thing and they need to. If, for instance, if the mix needs eighth notes, but they're going [sound effect] and it's not really pushing the mix, sometimes playing with compression on the acoustics or auditioning different kinds of compression to make it sound like, "Boy this guy was into it." Maybe pushing and pulling different instruments. Somebody's got to be back and sometimes it's better when things are back and other things are further up front. It's just basically playing with it and trying to put into it that un-definable thing that makes it exciting. Sometimes it means making sure your cymbals or your room mics are where you can actually feel the guy, or sometimes adding compression can be the answer to making the thing come alive. Sometimes hearing the guy breathe like the old Steve Miller records did. They had that [breathing]. With a little of that, you might say, "Man, he's working. I believe it." It's a little subconscious thing, but sometimes that can help.*

Like most other creative work that requires some divine inspiration for success, you can't underestimate talent and experience.

# CHAPTER 9

# Monitoring

A mixer is dependent upon his monitoring conditions and methods more than just about any other parameter. If the monitors don't work with the environment or if the mixer doesn't interact well with the monitors, then all the other tips and techniques are for naught.

## BASIC MONITOR SETUP

One thing frequently overlooked when auditioning near-field monitors is how the monitors are placed. This can make an enormous difference in the frequency balance and stereo field and should be addressed before you get into any serious listening. Here are a few things to experiment with before you settle on the exact placement.

### Check the Distance Between the Monitors

If the monitors are too close together, the stereo field will be smeared with no clear spatial definition. If the monitors are too far apart, the focal point or "sweet spot" will be too far behind you and you'll hear the left or the right side but not both together. A rule of thumb is that the speakers should be as far apart as the distance from the listening position. That is, if you are four feet away from the monitors, then start by moving them four feet apart so that you make an equilateral triangle between you and the two monitors. A simple tape measure will work fine to get it close. You can adjust them either in or out from there.

### Check the Angle of the Monitors

Improper angling will once again cause smearing of the stereo field as evidenced by a lack of instrument definition. The correct angle is determined strictly by taste, with some mixers preferring the monitors to be angled directly at their mixing position while others prefer the focal point (the point where the sound from the tweeters converges) anywhere from three to five feet behind them to eliminate some of the "hype" of the speakers.

To set the angle of the monitors, set up your monitors in the equilateral triangle fashion first, as described above. A great trick for getting excellent left/right imaging is to mount a mirror over each tweeter and adjust speakers so that your face is clearly seen in both mirrors at the same time when you are in your mixing position.

### Check How the Monitors are Mounted

Monitors that are mounted directly on top of a console meter bridge without any de-coupling are subject to comb filter effects, especially in the low end. That is, the sound travels through the console, through the floor and reaches your ears first (because it is denser material and travels faster) before the direct sound from the monitors through the air, causing phase cancellation. This can be more or less severe depending if the speakers are mounted directly on the metal meter bridge or mounted on a piece of carpet or similar material covering the metal meter bridge (very popular). The best way to de-couple the monitors is to use the same method used when soffit-mounting your main monitors. Set the near-fields on a half-inch or three-quarter-inch piece of open cell neoprene (soft rubber) and de-coupling will no longer be an issue.

Instead of mounting the near-fields on the console, a better solution is to mount them on stands just directly behind the meter bridge. Not only will this improve the low frequency de-coupling, but it will greatly decrease the unwanted reflections off the console.

### Check the Position of the Tweeters

Most mixers prefer that the tweeters of a two- or three-way system be on the outside, thereby widening the stereo field. Occasionally, tweeters to the inside works, but this usually results in smearing of the stereo image. Experiment with both, however, because you never know.

**Check the Console Itself**

The angle of the console, type of materials used for the panels, knobs, and switches, the type of paint and the size and composition of the armrest all make a difference in the sound due to reflections causing phase cancellation. If the sound of the near-fields on top of the meter bridge is unacceptable, then try moving them towards you with extenders or put them on stands behind the console (don't forget to de-couple them).

## HOW LOUD (OR SOFT) SHOULD IT BE?

One of the greatest misconceptions about music mixers (especially the great ones) is that they mix at high volume levels. In fact, quite the opposite is generally true. Most mixers find that they get better balances that translate well to the real listening world by monitoring at conversation level (79dB SPL) or lower.

High SPL levels for long periods of time are generally not recommended for the following reasons:

1) First, the obvious one: exposure to high volume levels for long periods of time may cause long term physical damage.

2) High volume levels for long periods of time will not only cause the onset of ear fatigue, but general physical fatigue as well. This means that you might effectively only be able to work 6 hours instead of the normal 8 (or 10 or 12) that are possible with lower levels.

3) The ear has different frequency response curves (remember the Fletcher-Munson curves?) at high volume levels that over-compensate on both the high and low frequencies. This means that your high volume mix will generally sound pretty limp when played at softer levels.

4) Balances tend to blur at higher levels. What tends to work at higher levels won't necessarily work when played softer. However, balances that are made at softer levels always work when played louder.

Now this isn't to say that all mixing should be done at the same level and it should all be soft. In fact, music mixers (as opposed to film, which always has a constant SPL level) tend to work at a variety of levels: up loud for a minute to check the low end, moderate while checking the EQ and effects. But the final balances nearly always will be done quietly.

**Don Smith:** *I like to listen loud on the big speakers to get started, and occasionally thereafter, and most of the time at about 90dB. When the mix starts to come together, it comes way down, sometimes barely audible. I turn it down way low and walk around the room to hear everything.*

**Allen Sides:** *Generally speaking when I put up the mix, I'll put it up at a fairly good level, maybe 105, and set all my track levels and get it punchy and fun sounding. And if I listen loud, it's only for very short periods of time. It's rare that I would ever play a track from beginning to end loud. I might listen to 20 seconds or 30 seconds of it here and there, but when I'm actually down to really detailing the balance, I'll monitor at a very modest level. I would say at a level that we could have a conversation and you could hear every word I said.*

**Ed Seay:** *I mix at different levels. I try not to mix too loud because it'll wear you down and fool your perspective. I don't find it extremely valuable to listen loud on big wall monitors very often. The only reason I'll go up there is to check bottom end.*

*Sometimes it's very valuable to turn things down, but there's an up and down side to both. If you listen too soft, you'll add too much bass. If you listen too loud, you'll turn the lead vocals down too much. What I like to do is make it sound good on all three unrelated systems, then it's got to relate to the rest of the world.*

**George Massenburg:** *I'll monitor way loud to see what rocks. I'll monitor at a nominal level to get sounds together. Then I'll monitor about 5dB over background noise to hear all the elements into focus. If a mix works at 30dB SPL, 25dB SPL, it'll almost always work a lot louder. If you can hear everything at that low a level, then when you turn it up you'll have a very even balance. That's the way to get everything in the same plane, by listening extremely low.*

**Guy Snider:** *I monitor extremely soft, to the point that assistants have to leave the room. I have a tendency to pick up nuances at conversation volume. Like if you were in the room talking for any length of time, I'd either ask you to leave or I'd turn up the volume all the way in the mains until it shut you up because I can't have conversation in the room while I mix.*

**Jon Gass:** *Like the SSL up on one (the Control Room Monitor level control) is what I mix on most of the time. It's really quiet but I can mix long and not get fatigued. Sure, I do the NS10 thing, and then towards the end of the mix I'll go really loud on the NS10's and do some adjusting, and I'll go extremely loud on the big ones and do some more adjusting just to fine-tune.*

**David Pensado:** *I usually listen to NS10's kind of medium and Auratones I listen at the same volume you would listen to TV. I found that on the NS10's, in order for them to really work, it's best to have them stay at one level for most of the mix. Then near the end of the mix, check your levels and your EQ with the NS10's about 20 percent lower and again about 20 percent higher and you'll make adjustments that you'll really be pleased with when you hear it on the radio. The big speakers I use mostly to show off for clients and to just have fun. I like to turn it up and, if my body is vibrating properly, then I'm happy with the low end. A lot of engineers use them to hype the client, but I also use them to hype myself! If I'm cranking and I'm not getting excited, then I just keep on working.*

**Ken Hahn:** *I personally monitor about as low as most people would accept. I tend to go that way because inevitably, if you get it sounding good at a low level, it just sounds that much better at higher levels. It sort of forces you to do a lot more manual gain riding at low level because otherwise stuff just doesn't poke through. I'm sort of doing my own form of manual compression and I've found that usually works better than the other way around.*

**David Sussman:** *On the Yamahas I'll listen at a low level like 2 or 3. Then I'll graduate to 6 or 7, slide my chair back from the board, and just try to get more of an out-of-the-image listen. Then I'll listen up top on the big speakers really loud, just to make sure that I've got the bottom right. Otherwise, I usually like to mix at a pretty relatively low volume for as long as I can.*

**Allen Sides:** *Yeah, there's also a question of dryness versus live-ness versus deadness in regards to monitor volume. Obviously, when you turn it down your ambience determines how loud it sounds to you to some degree. And if you're monitoring at a loud level and it's very dry, it can be very impressive sounding. When you turn down, it might not be quite so full sounding so obviously there's a balance there.*

Since sooner or later your mix will be played back in mono somewhere along the line, it's best to check what will happen before you're surprised later. Listening in mono is a time-tested operation that gives the mixer the ability to check several things:

- Phase Coherency
- Balances
- Panning

### Phase Coherency

When a stereo mix is combined into mono, any elements that are out of phase will drop in level or even completely cancel out. This could be because the left and right outputs are wired out of phase (pin 2 and pin 3 of the XLR connector are reversed), which is the worst case scenario, or perhaps because an out-of-phase effect causes the lead vocal or solo to disappear. In any event, it's prudent to listen in mono once in a while, just to make sure that a mono disaster isn't lurking in the wings.

### Balances

Many engineers listen to their mix in mono strictly to balance elements together since they feel that they hear the balance better this way. Listening in mono is also a great way to tell when an element is masking another.

**Joe Chiccarelli:** *I listen in mono an awful lot and find it's great for balances. You can easily tell if something's fighting.*

**Andy Johns:** *People don't listen in mono any more but that used to be the big test. It was harder to do and you had to be a bloody expert to make it work. In the old days we did mono mixes first, then did a quick one for stereo. We'd spend eight hours on the mono mix and half an hour on the stereo.*

### Panning

Although not many engineers are aware that their stereo panning can be improved while listening in mono, this is in fact a good way to achieve a level of precision not available in stereo.

**Don Smith:** *I check my panning in mono with one speaker, believe it or not. When you pan around in mono, all of a sudden you'll find that it's coming through now and you've found the space for it. If I want to find a place for the hi-hat for instance, sometimes I'll go to mono and pan it around and you'll find that it's really present all of a sudden, and that's the spot. When you start to pan around on all your drum mics in mono, you'll hear all the phase come together. When you go to stereo it makes things a lot better.*

## MONITORS — WHICH ONE?

So which speaker is best for you to monitor on? Certainly there's plenty of choices, and there is clearly no single favorite among the great mixers. Probably as close to a standard as we have today is the Yamaha NS10, closely followed by the Auratone. Auratones have fallen out of favor since their peak of popularity during the 70's, but most mixers still use them as an additional reference (although sometimes they use only one, in mono).

### Things to Listen for in a Monitor

**Even Frequency Balance** – While listening to a piece of music that you know well, check to see if any frequencies are exaggerated or attenuated. This is especially important in the crossover area (usually about 1.5 to 2.5kHz). Listen especially to cymbals on the high end, vocals and guitars for the midrange and bass and kick drum on the low end.

**Frequency Balance Stays the Same at Any Level** – The less the frequency response changes as the level changes (especially when playing softly), the better. In other words, the speaker should have roughly the same frequency balance when the level is soft as when it's loud.

**High Output Level Without Distortion** – Be sure that there's enough clean level for your needs. Many powered monitors have built-in limiters that stop the speaker or amplifier from distorting, but also may keep the system from getting as loud as you may find necessary.

The number of monitor references that are used is an important aspect to getting a mix right. Although a mixer may do most of his work on a single system, it's common to check the mix on at least two (maybe more) other sources as well. Usually this will be the main soffit-mounted monitors, the near-field monitors of choice (which may be NS10's) and an alternative which could be

Auratones, NS10's or just about anything else. Couple that with the ever-present boom box, car stereo or stereo in the lounge, and the average of all these systems should make for a good mix.

Until recently when the trend turned towards powered monitors, many engineers also brought their own amplifiers to the studio. This is because the amp/speaker combination is a delicate one, with each speaker having a much greater inter-dependence with the power source than most of us realize. In fact, the search for the perfect amplifier was almost as long-suffering as the search for the perfect monitor. All of this has dwindled in recent years, thanks to monitors with built-in a-mplifiers perfectly matched to its drivers.

**Jerry Finn:**   *When I was an assistant, a lot of the engineers that I liked working with had Tannoy SRM10B's. When I went independent, I searched high and low and finally found a pair. I carry those around with me wherever I go as well as a Hafler Transnova amp, which gets frowned upon some-times amongst the guys that are into the more hi-fi kind of thing. But I tried 20 amps and that just sounded the best.*

Even though NS10's reside in just about every studio, there are two camps, consisting of lovers who wouldn't mix without them and haters who never touch them. It would certainly be folly to use these speakers just because everyone else is (because they're not all using them, for one thing). In fact, it's not a good idea to use *any* particular speakers unless you're really in love with them. You'll have to listen to these monitors for a lot of hours so you might as well like what you hear.

In my frequent speaker auditions for *EQ Magazine* over the course of five years, I've found that you can easily get used to just about any speaker if you use it enough and learn its strengths and weaknesses. It also helps to have a very familiar, solid reference to compare the sound. For instance, if you know how things sound in your car, then adjust your mixes so they work when you play them there. I usually use mastering engineer Eddy Schreyer at Oasis Mastering as my reference. When I do a mix on a new set of speakers he'll tell me, "You're off a dB at 5k, a dB and a half at 150 and -2 at 40." I'll then adjust accordingly.

**George Massenburg:** *I'm a big one for hallway. I hate cars. Through the control room doors is always an important thing for me, because I almost never do loud playbacks. I like listening around the corner and on a blaster.*

**Guy Snider:** *That's one thing that's really improved my mixing in the last three years: when I started gearing my mixes towards the boombox, and I quit taking my Tannoys to the studio and started using the old Yamaha NS10N's for everything except getting my drum sounds. All of a sudden, the boombox tape started sounding better and better and better and better.*

**Joe Chiccarelli:** *I'll walk out of the control room and listen to it right outside the door. It's interesting to hear what it sounds like through the crack in the door. Things pop out. …Blasters are good things for sure as well.*

*The Mixing Engineer's Handbook*

Most everyone wants to hear what their mix sour
different speakers and in different environments
consumer oriented perspective. Here's some of t
standards:

> The car
> Boombox
> Through the control room door

**Jon Gass:** *One of my favorite ones is to turn on the vacuum clean*
*against the wall in the front of the room. Sounds a little*
*just kind of want to see if the mix is still cutting throug*

**Andy Johns:** *Obviously, the idea is to make it work on all systems. Yo*
*big speakers, the NS10's, out in the car, plus your own s*
*you go home and listen again. This is a lot of work but*
*to go.*

**David Sussman:** *I have one particular room that I mix a lot at, Studio B*
*Studios. I usually lay down with my head on the armrest*
*the back wall when I'm checking my bottom in that room.*
*the big speakers up to like 8, it has to hit me a certain wa*
*then I know something's not right.*

**Don Smith:** *I mix a lot at my house now where I sit outside a lot on n*
*mix in a studio with a lounge I'll go in there with the con*
*shut and listen like that. I definitely get away from the mi*
*speakers as much as possible.*

**Ed Seay:** *What I'll do about an hour before printing the mix is prop*
*open the control room door and walk down the hall or int*
*the lounge where the music has to wind its way out the do*
*I find that very valuable because it's not like hitting mono*
*console, it's like a true acoustic mono. It's real valuable to*
*all the parts and it's real easy to be objective when you're n*
*the speakers and looking at the meters.*

Good ad producers use a similar technique: flip to a
mono Auratone, lower the volume to just perceptibl
see if it still sounds like a commercial. Then raise the
a tiny bit, walk out into the hall and see if you still lik

# CHAPTER 10

# The Master Mix

G one are the days of manual mixing, where the hands of not only the engineer but the producer and all of the band members manned a fader or a mute button or a pan control in order to get the perfect mix. Gone are the days of massive numbers of takes in order to make sure that you had the last best one before you got your "keeper." Thanks to the advanced state of console automation, the mix is perfect before it ever gets committed to tape, DAT, MO (Magneto Optical), or hard disk.

Just what is everyone mixing to nowadays, anyway?

## MIX-DOWN FORMATS

Although this might change soon, many mixers still prefer to mix to the now old fashioned 1/2" two-track analog tape at 30ips and information from the major mastering houses indicate a 50/50 split between this format and Digital Audio Tape (DAT).

Half-inch analog is still attractive for several reasons:

• **The sound, of course** — Even though mixing engineers tend not to agree on almost anything, most like the sound of analog over other formats. Many mixers run both analog and DAT at the same time and choose which one is best for the particular song, but the DAT choices are usually in the minority.

This may change in coming years as higher sampling rates, larger bit depths and better converters become more commonplace in the digital domain.

- **Archival purposes** — While many distrust the longevity of the various digital formats, it's almost universally agreed that analog will be able to maintain its integrity for enough time until a new replacement is found. After all, tapes made in the early 50's still play back and even sound as good if not better than current examples.

  Also, with sample rates and bit depths going up all the time, many feel that an analog master can fulfill the needs of the future better than a DAT that is limited to the current restricted 44.1kHz sample rate and 16-bit standard.

- **The cost** — Although 1/2" analog costs more than DAT, using it is less costly these days thanks to console automation. Mixing to analog used to be an expensive proposition just in the matter of tape costs. In the days prior to automation it would be commonplace to use upwards of 50 reels to mix an album due to the multiple takes that it took the mixer to get all the level and mute moves manually flawless. But now with consoles so thoroughly automated, a mixer's every move is remembered and will be committed to tape only when the mix is deemed perfect, saving tremendously on tape costs from unneeded and unwanted mix versions.

  DAT, on the other hand:

- Is **way cheaper than analog tape**. A two-hour DAT costs only $10 (or less) while a reel of 1/2" analog tape that will store only 16.5 minutes at 30ips costs about $40.

- If used with high quality outboard analog to digital and digital to analog converters, DAT **can sound very good indeed**. The real key is those outboard converters, though. The normal onboard converters of just about any DAT machine yields quality far below what can be had from their better external counterparts. Plus, with the introduction of the new 24-bit models, DAT machines sound better than ever.

- Because of its size, DAT is far **easier to transport, store and mail**. But, because of the size again, it's much more difficult to add the identifying information that's sometimes required for a master tape.

Two new alternatives have come on the scene the last few years: Magneto Optical (MO) and hard disk recording via a Digital Audio Workstation (DAW). Since most mixes are eventually loaded onto an audio workstation for editing, it makes sense that the mixes are recorded directly onto the workstation (like Protools or Sonic Solutions) instead of onto tape or DAT first. This, of course, eliminates a step but also eliminates a source of backup should anything corrupt the data. Usually in these situations, a DAT is run as a backup as well.

## ALTERNATIVE MIXES

In these days of automation on nearly every console, it's become standard operating procedure to do multiple mixes in order to avoid having to redo the mix again at a later time because an element was mixed too loud or soft. This means that any element that might later be questioned, such as lead vocal, solo, background vocals and any other major part, is mixed with that part recorded slightly louder and again slightly softer. These are referred to as the *up mix* and the *down mix*. Usually these increments are very small, 1/2 to 1dB, but never more.
With multiple mixes, it's also possible to correct an otherwise perfect mix later by editing in a masked word or a chorus with louder background vocals.

Thanks to the virtues of modern automation, many engineers leave the up and down mixes to the assistants since most of the hard work is already done.

**Allen Sides:** *Invariably I will do the vocal mix to where I'm totally happy with it and then I'll probably do a quarter and half dB up and a quarter and half dB down. I really cover myself on mixes these days. I just do not want to have to do a mix again.*

**Benny Faccone:** *Usually one with the vocal up .8dB and another with the vocal down .4dB, and if there's backgrounds, the same thing. I do not want to come back to remix. Once I'm done with a song I've heard it so much that I don't want to hear it ever again.*

**Don Smith:** *I try to just do one mix that everybody likes and then I'll leave and tell the assistant to do a vocal up and vocal down and all the other versions that they might want, which usually just sit on a shelf. I'll always have a vocal up and down versions done because I don't feel like remixing a song once it's done.*

**Ed Seay:** *Generally I like to put down the mix and then I'll put down a safety of the mix in case there was a dropout or something went goofy that no one caught. Once I get the mix, then I'll put the lead vocal up half a dB or 8/10 of a dB and this becomes the vocal up mix. Then I'll do a mix with all vocals up. Sometimes I'll recall the mix and just do backgrounds up and leave the lead vocals alone. Then I'll do one with no lead vocal and just the backgrounds. Then I'll do one with track only, just instruments. That's usually all the versions I'll need to do.*

**Ed Stasium:** *I'll do a vocal up. Sometimes I do guitars up. You get so critical when you're mixing and when it comes down to it, it's the darn song anyway. As long as the vocal's up there, it will sound pretty good. You won't even notice the little things a month later.*

**Joe Chiccarelli:** *I'm really bad about that because I'll do a lot of options. I'll always do a vocal up in case someone at the record company complains that they can't hear a line. I'll always do a bass up or even a bass down as well. When I say up, I'm talking about a quarter or half dB because I find that if you get your balances good enough, that's the only amount of alteration you can make without throwing everything totally out of whack. A lot of times I'll do a number of other options like more guitar, more backgrounds, or whatever key element that someone might be worried about. And then sometimes if I'm not feeling like I got the overall thing right, I might do one more version that has a little tweak on that as well. Sometimes I'll add like a Massenburg EQ on the stereo buss and add a little 15k and maybe some 50 as well to give the record a little more of a finished master sound.*

**Jon Gass:** *I'll do the main version, a lead up, just the backgrounds up and then the lead and backgrounds up. I hardly ever do a vocal down version. Then I'll just go through and pick some instruments that somebody might have questioned, "Is that loud enough?" I'll do those kind of things. It usually comes out to be 10 or 12 versions of each song, believe it or not.*

**Lee DeCarlo:** *I do a lot. I like to play around with it. I have always thought it would be a wonderful thing to mix your entire album in a day. And instead of doing one song a day for ten days, it would be a really great idea to mix the entire album ten times; then go back and listen to which ones you like the best.*

In extreme cases, some mixers have resorted to the use of "stems" in order to keep everyone (mostly the record company) happy. A stems mix is usually done on an 8-track MDM such as a DA-88 or ADAT and consists of a stereo bed track and individual stereo tracks of the most important elements complete with effects. This allows for an easy remix later if it's decided that the balance of the lead vocal or the solo is wrong.

Stems are widely used in film mixing because a music mixer usually cannot tell what's going to be too loud or be masked by the additional dialogue or sound effects elements. The stem mix gives the dubbing mixer more control during the final mix if required.

# Part II
## Mixing in Surround

# CHAPTER 11

# Mixing in Surround

W hile most of this book applies to mixing in stereo, music mixers will soon be faced with a new task; mixing for surround. Surround sound is almost universally acclaimed to be a more realistic and pleasing experience to the listener than stereo. This applies to just about any type of program, from music to motion pictures to television. People that can't tell the difference between mono and stereo can immediately hear and appreciate the difference between surround and stereo. It is a development so dramatic that it will change the way we listen, record, mix and enjoy music forever.

## A BIT OF HISTORY

Surround sound in one form or another has actually been with us for more than 50 years. Film has always used the three channel "curtain of sound" (developed by Bell Labs in the early 30's) since it was discovered that a center channel provided the significant benefits of anchoring the center by eliminating "phantom" images (in stereo the center images shift as you move around the room) and better frequency response matching across the sound field. The addition of a rear effects channel to the front three channels dates back as far as 1941 with the "Fantasound" four channel system utilized by Disney for the film *Fantasia* and in the 1950's with Fox's Cinemascope, but it didn't come into widespread use until the 60's when Dolby Stereo became the surround de facto standard. This popular film format uses four channels (left, center, right and a mono surround, sometimes called *LCRS*) and is encoded onto two tracks. Almost all major shows and films currently produced for theatrical release and broadcast television are presented in Dolby Stereo since it has the added advantage of playing back properly in stereo or mono if no decoder is present.

With the advent in the 80's of digital delivery formats capable of supplying more channels, the number of surround channels was increased to two and the Low Frequency Effects channel was added to make up the six-channel 5.1 system, which soon became the modern standard for most films (the Sony SDDS 7.1 system being the exception), music and DTV.

And of course, there's Quad from the 70's, the music industry's attempt at multi-channel music that killed itself as a result of two non-compatible competing systems (a preview of the Beta vs. VHS war) and a poor psychoacoustic rendering that suffered from an extremely small sweet spot.

## TYPES OF SURROUND SOUND

5.1 is the mostly widely used surround format today, being used in motion picture, music and digital television. The format consists of six discrete speaker sources: three across the front (left, center and right), two in the rear (left surround, right surround) plus a sub-woofer (known as the Low Frequency Effects channel or LFE), which is the ".1" of the 5.1 (see Figure 13). This is the same configuration that you hear in most movie theaters, since 5.1 is the speaker spec used not only by THX but also by popular motion picture release formats such as Dolby Digital and DTS.

**Figure 13  A 5.1 Surround System**

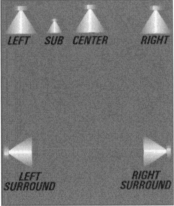

Graphic courtesy of Dolby Labs

### The LFE Channel

LFE stands for Low Frequency Effects and is sometimes referred to in film production circles as the "boom" channel because that's what it's there for: to enhance the low frequencies of a film so you get the extra boom out of an earthquake, plane crash, explosion or other such dramatic scene requiring lots of low frequencies.

The LFE channel, which has a frequency response from about 25Hz to 120Hz, is unique in that it has an additional 10dB of headroom built into it. This is needed to accommodate the extra power required to reproduce the low frequency content without distortion.

## BASS MANAGEMENT

The Bass Manager (sometimes called Bass Redirection) is a circuit that takes all the frequencies below 80Hz from the main channels and the signal from the LFE channel and mixes them together into the subwoofer. This is done to make use of the subwoofer for more than the occasional low frequency effect, since it's in the system already. This enables the effective response of the system to be lowered to about 25Hz.

Since the overwhelming majority of consumer surround systems (especially the average low end ones) contain a bass management circuit, if you don't mix with one then you're not hearing things the way the people at home are. And, since the Bass Manager gives a low frequency extension below that of the vast majority of studio monitors, the people at home may actually be hearing things (like unwanted rumbles) that you can't hear while mixing.

## OTHER TYPES OF SURROUND

There are many other widely used surround formats. Three-channel (stereo front speakers with a mono surround), four-channel (three front speakers with a mono surround) such as Dolby Prologic, five-channel (three front speakers with a stereo surround but no LFE channel), and eight-channel (the Sony SDDS format with five front speakers and an LFE channel) all abound.

There are other non-standard formats that use as many as ten channels for height and extra rear and side channels as well. The Star Wars prequel *Episode 1—The Phantom Menace* introduces a Dolby Digital Surround EX 6.1 format in which a center rear channel is used. Many amusement rides such as Universal's *Back to the Future* ride have used as many as 14 channels.

# CHAPTER 12

# Why Is Surround Better Than Stereo (or Quad for that Matter)?

hen you listen to surround sound you'll notice quite a few improvements over stereo:

- The **sonic clarity** is enhanced because the center channel anchors the sound and eliminates any phantom image shifts that we take for granted in stereo.

- **There is no sweet spot per se.** — Actually, the whole room becomes a sweet spot in that you can move around freely and never loose the sense of clarity, dimension and spatial continuity. One listener described it perfectly as an "audio sculpture" in that, just like when you walk around a piece of artwork and get a different perspective of the art, when you walk around the 5.1 room you just get a different perspective of the mix. You might get closer to the guitar player, for instance, if you walk to the left of the room. Walk to the right and you're closer to the piano. Indeed, you don't have to even be in the speaker field to get a sense of the depth of the mix. While mixing, people sitting on a couch outside of the soundscape often describe an enhanced experience.

- **Speaker placement is very forgiving.** — Yes, there are standards for placement, but these tend to be very non-critical. The sense of spaciousness remains the same regardless of how haphazardly the speakers are distributed around the room. In fact, stereo is far more critical placement-wise than surround sound.

## SURROUND MIXING

During mixing, there are several surprising advantages:

- **Clarity of instruments** — Everything sounds much more distinct as a result of having more places to sit space-wise in the mix. This means that you spend a lot less time EQing, trying to get each instrument heard.

- **Added dimension** — Even mono tracks are big and dimensional in surround! No longer is there a need to "stereo-ize" a track by adding an effect. Simply spreading a mono source across the speakers with the pan pot makes it big sounding.

- **The ambience is different** — When you mix in stereo, usually you must recreate depth. In surround, it's built-in. Because of the naturally increased clarity and dimension, you no longer have to spend as much time trying to artificially add space with reverb, delays, etc. This is not to say that you won't use these effects at all, but the approach is different, since surround automatically gives you the depth that you must artificially create with stereo.

- **Mixes go faster** — It actually takes less time to do a mix because surround sound automatically has a depth of field that you normally have to work hard to create when you're mixing in stereo. Most mixers find they need less EQ and less effects because there's more room in the soundscape to place things.

### Differences Between Surround for Picture and for Music

Normally in the theater, all of the primary sound information comes from the front speakers and the surround speakers are utilized only for ambience info in order to keep your attention on the screen. The LFE is intended to be used just for special effects like explosions and earthquakes and is therefore used infrequently. One of the reasons that the surround speakers don't contain more source information is a phenomena known as the "exit sign effect," which means that your attention is drawn away from the screen to the exit sign when the information from the surrounds is too loud.

But music-only surround sound has no screen on which to focus and therefore no exit sign effect to worry about. Take away the screen and it becomes possible to utilize the surround speakers for more creative purposes.

### Classical vs. "Middle of the Band"

There are two schools of thought about how surround sound for music should be mixed. The Classical method puts the music in the front speakers and the hall ambience in the surrounds, just as if you were sitting in the audience of a club or concert. This method may not utilize the LFE channel at all and is meant to reproduce an audience perspective of the musical experience.

In the case of the Middle of the Band method, the band is spread all over the room via the five main speakers (the LFE may be used for bass and kick, which is also spread to the other speakers as well) and that puts the listener in the center of the band and envelopes him with sound. This method usually results in a much more dramatic soundstage that is far bigger sounding than the stereo that we're used to. This may not be as authentic a soundscape as some music might require, however (for example, any kind of live music where the listener's perspective is from the audience).

## WHAT DO I PUT IN THE CENTER CHANNEL?

In film mixing, the center channel is used primarily for dialogue so sonic movement doesn't distract the listener. In music, however, its use prompts debate among mixers.

### No Center Channel

Many veteran engineers who have mixed in stereo all their lives have trouble breaking the stereo paradigm to make use of the center channel. These mixers continue to use a phantom center from the left and right front speakers and prefer to use the center speaker as a height channel or not use it at all.

### Isolated Elements in the Center Channel

Many mixers prefer to use the center channel to isolate certain elements such as lead vocals, solos and bass. While this might work in some cases, many times the isolated elements seem disconnected from the rest of the soundscape (see Figure 14).

**Figure 14  Isolated Elements in the Center Channel**

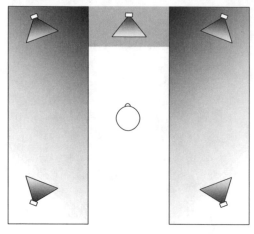

**The Center as Part of the Whole**

Mixers who use the center channel to its fullest find that it acts to anchor the sound and eliminates any drifting phantom images. In this case, all five speakers have equal importance with the balance changing the sound elements placed in the sound-scape. It's actually easiest to picture this as in Figure 15, with the soundscape cut in half from the middle of the center speaker.

**Figure 15  Integrated Center Channel**

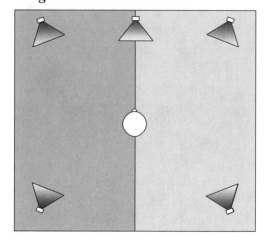

## WHAT DO I SEND TO THE LFE (SUB-WOOFER) CHANNEL?

Anything that requires some low frequency bass extension can be put into the sub-woofer via the LFE channel. Many mixers put a little kick and/or bass there if it's used at all. Remember that the frequency response only goes up to 120Hz so you have to put the instrument into the main channels as well, in order to gain some definition.

In fact, it might be better not to use the LFE channel unless you're positive that the sub-woofer is calibrated correctly. An un-calibrated sub-woofer can cause big surprises in the low end when the track is later played back on the typical home theater setup. If you don't use the sub when mixing, the low frequencies under 80Hz are naturally folded into the playback sub-woofer resulting in a smooth and even response.

## SURROUND TO STEREO COMPATIBILITY

Although it's possible to have the surround mix automatically down-mixed to stereo either via SMART Content down-mixing inherent in a DVD-Audio disc or by selection of the down-mix parameters on the Dolby Digital encoder, the results are often unpredictable and many times unsatisfactory. It's best to prepare a separate dedicated stereo mix whenever possible since their will most likely be sufficient room on the delivery medium (DVD).

## SURROUND MASTER RECORDERS

Although any multitrack format can be used as a master recorder, the de facto standard is the Tascam DA-88 family (DA-98, PCM800, etc), sometimes enhanced to 20-bit resolution with either Rane or Prism bit splitters. Other machines being used include the Genex GX8000 and 8500 Magneto Optical recorders and Tascam MMR-8 hard disk recorder. Some people are even mixing to 1- or 2-inch 8-track analog.

## MASTER TAPE TRACK ASSIGNMENTS

Sooner or later during one's first surround mix, the questions of channel assignment on the master recorder (be it tape or disc) always arise. What is the correct track assignment? Actually, there are several generally accepted channel assignment formats for surround, although the first is fast becoming the de facto standard. That is:

| Channel 1 | Channel 2 | Channel 3 | Channel 4 | Channel 5 | Channel 6 |
|-----------|-----------|-----------|-----------|-----------|-----------|
| Left Front | Right Front | Center | LFE | Left Surround | Right Surround |

A dedicated stereo mix, or Lt, Rt or encoded AC3 can be recorded onto Tracks 7 and 8. This format transfers easily to the corresponding four audio tracks (L, R, C, LFE) of the video formats widely used today such as DigiBeta or D5. This accommodates the necessary L, C and R tracks, as well as the common pairing of channels in Dolby Digital. The surround products of Panasonic, Mackie and Martinsound, to name just a few, now support this configuration. It is also the SMPTE and ITU standard.

The following two assignment methods are also used, but less and less as the above takes hold:

### FILM STYLE

| Channel 1 | Channel 2 | Channel 3 | Channel 4 | Channel 5 | Channel 6 |
|-----------|-----------|-----------|-----------|-----------|-----------|
| Left Front | Center | Right Front | Left Surround | Right Surround | LFE |

### DTS

| Channel 1 | Channel 2 | Channel 3 | Channel 4 | Channel 5 | Channel 6 |
|-----------|-----------|-----------|-----------|-----------|-----------|
| Left Front | Right Front | Left Surround | Right Surround | Center | LFE |

The above assignment is preferred by DTS. Again, the pairings are logical, but the placement is different from the Dolby standard. Tracks 7 and 8 usually contain the stereo version of the mix, if one is needed.

# CHAPTER 13

# Data Compression

D ata compression is the process of using psychoacoustic principles to reduce the number of bits required to represent the signal. This is needed with surround sound so more data can be squeezed onto a finite storage space, such as a CD or DVD, and also because the bit rate of six channels of 96/24 LPCM is too large to fit through the small data pipe of a DVD.

## HIGH OR LOW RESOLUTION – 96/24 VS. 48/20

First, understand that the first number (96) represents the sample rate in 1000 per second increments, or a sample rate of 96kHz. The second number (24) represents the word length of the encoded digital data, or 24 bits.

In order to understand the significance of each parameter and how it affects quality, a brief discussion of sampling rate and word length is in order.

The analog audio waveform is measured in amplitude at discrete points in time, and this is called *sampling*. The more samples of the waveform that are taken, the better the representation of the waveform, with a greater resultant bandwidth for the signal. Audio on a CD has a sampling rate of 44,100 times a second (44.1kHz) which yields a bandwidth of about 22kHz. A sampling rate of 96kHz gives a better digital representation of the waveform and yields a usable audio bandwidth of about 48kHz. Therefore, the higher the sampling rate, the better the representation of the signal and the greater the bandwidth.

The more bits a word has, the better the dynamic range. Every bit means 6dB in dynamic range. Therefore, 16-bit yields a maximum dynamic range of 96dB, 20-bit equals 120dB DR, and 24-bit gives a theoretical maximum of 144dB DR. There are no true 24-bit systems yet.

From this you can see that the high-resolution 96/24 format is far closer to sonic realism than the current CD standard of 44.1/16. The higher the sample rate, the greater the bandwidth, and therefore the better the sound. The longer the word length (more bits), the greater the dynamic range, and therefore the better the sound.

What all this means is that the mixer now has the choice of which resolution to mix to, which was never available before. For the highest fidelity, a stereo mix at 192/24 can be chosen, or 96/24 for 5.1 surround. It's also possible to choose any number of other possibilities such as 96/24 for the front channels and 48/16 for the rear or 48/20 for all channels.

Thanks to the new DVD medium, mixers are no longer tied to the old CD quality standard of 44.1kHz at 16 bits.

## DOLBY DIGITAL (AC-3) OR DTS ENCODING

Both Dolby Digital® and DTS (Digital Theater Systems) are what are known as lossy data compression schemes where some information that is masked by more prominent data is thrown away. This is done in order to fit a lot of data through a small data pipe. Dolby Digital® (sometimes called AC-3, which is the perpetual coding system used in Dolby Digital) takes six channels of 48kHz/24-bit information and compresses it at about an 11:1 ratio to a maximum bit rate of 640kbps, although 384 is the average data rate used. DTS compresses at about a 3:1 ratio at an average data rate of 1.4Mbps. Because there is less data compression, many like the sound of DTS encoded product better.

### Lossy and Lossless Compression

Lossy compression (such as Dolby Digital® or DTS) is built around perceptional algorithms that remove signal data that is being masked or covered up by other signal data that is louder. Because this data is thrown away and never retrieved, it's what's known as *lossy*.

Depending upon the source material, lossy compression can either be completely inaudible, or somewhat noticeable. It should be noted that even when it is audible, lossy compression still does a remarkable job of recovering the audio signal and still sounds quite good.

Lossless compression, such as MLP (see below), never discards any data and recovers it completely during decoding and playback.

### Meridian Lossless Packing

Meridian Lossless Packing is the compression standard used on the DVD-Audio disc in order to store six channels of high resolution 96/24 audio. MLP's main feature is that it never discards any signal information during data compression (which is why it's *lossless*) and therefore doesn't affect the audio quality. MLP gives a compression ratio of about 1.85:1 (about 45 percent) and its licensing is administrated by Dolby Laboratories.

## SURROUND ENCODERS

Although it's not imperative that an encoder be present at the mix, it does help to hear what the codec (the Dolby, DTS or MLP compressor/decompressor) will do to the final product because codecs can change the sound considerably. There are also quite a few parameters (like down-mixing and Late Night Levels) that the producer might like to tweak rather than leave for someone else down the production chain.

Down-mixing automatically folds down the 5.1 surround program to the available number of channels. In other words, if only two speakers are available, then the surround mix is folded down to stereo. Although this is less desirable than a separate mix, the Dolby Digital® encoder gives a number of choices about how this is done.

Late Night Levels is essentially the same thing as a Loudness control for surround sound. When a surround system is calibrated, it's usually done at a fairly moderate level of 85dB SPL. This level is usually way too loud for quiet listening late at night so the system naturally gets turned down, which may destroy the balance between the front and rear speakers and the sub-woofer.

The Late Night parameter takes this into account and allows the mixer to somewhat compensate for the balance discrepancies.

For more information of Surround Sound production, delivery methods, and calibration, visit the Surround Sound FAQ at: **http://www.surroundassociates.com/safaq.html**

## SURROUND MASTER MEDIA PREP

Surround sound brings a whole new level of complexity not normally found in stereo. Therefore, it's imperative that you indicate as much information about your project as possible. Many potential problems can be avoided as long as the master is prepped and the following items are noted.

### Slate the Master

More than ever before, it's important to not only properly document the master tape or disc, but also to prep the master in order to be sure that there are no questions as to the actual track assignments. Even an engineer who's mixed the tracks sometimes has a hard time determining which is the center and which is the left surround. Thus, it's quite necessary to take the guesswork out of the process.

The best way to avoid confusion is to go back to the admittedly low-tech but foolproof method of using an audio slate on each channel indicating the channel assignment (e.g., "Channel One — Left Front," Channel Six — Right Surround").

### Print a Test Tone

Print at least 30 seconds of 1kHz tone at −20dBFS, which is the SMPTE standard reference level, across all tracks. A 1k tone is a pretty good way to discover if there are any clock discrepancies since the purity of the signal will suffer as a result of the clicks and warbles which might not be heard during the actual program material.

Also keep in mind that any program on tape media should start at no earlier than two minutes into the tape, since that's where most errors and dropouts usually occur.

## Other Things That Should Be Documented

• Is the LFE channel filtered and at what frequency?

This is important if for no other reason than it's easy to figure out which is the sub-woofer channel if the assignment documentation is lost.

• What is the reference level in SPL?

This helps the mastering engineer to better approximate what you were hearing during the mix if there is a problem down the line.

• What is the sampling rate?

This helps to avoid any clock or synch issues that may arise during the mastering or authoring.

• What is the bit resolution?

This is necessary in order to set dither correctly. Dither is a small noise signal of only a dB or so that's intentionally induced into the digital world in order to remove unused or unwanted bits at the end of the digital word (e.g., 24 bit program must be converted to a 20 bit program). To simply lop off (truncate) the bits at the end of the word sounds bad, so dither is used instead.

• If time code is included, what is the format?

If the audio program is linked to picture, the time code format is necessary to maintain sync.

• Are the surround channels calibrated equal to the front channels or −3dB?

In film style mixing, the surround channels are calibrated at −3dB. Music style mixing has the surrounds equal in level to the front speakers.

• What is the media format and how many pieces are there?

The entire master may be on several pieces of media across several different formats. A warning here can eliminate the confusion of an incomplete mastering or authoring job later.

• How long is the program?

This is necessary because it determines if data compression must be used.

• What is the intended final audio resolution (e.g., 96/24 or 48/20)?

Once again, this determines if data compression is used and how it is set.

• Any glitches, distortion, dropouts or anything unusual?

Good things to indicate on any master tape, this stops the mastering engineer from checking his own equipment when a defect lies in the master.

# Part III
## Interviews

# Joe Chiccarelli

E ven though he may not have quite as high a profile as many other notable big-time mixers, engineer/ producer Joe Chiccarelli's list of projects are equally as notable as the best of the best. With credits like Tori Amos, Etta James, Beck, U2, Oingo Boingo, Shawn Colvin, Frank Zappa, Bob Seger, Brian Setzer, Hole and many, many more, chances are you've heard Joe 's work more times than you know.

*How long does it take you to mix a track?*
It really depends on the material, the amount of tracks, and the arrangement. I try to work fast because I find that the longer it takes the more I get into a sort of myopic mindset and get bogged down with the little details. You miss the vibe and the big picture and just suck the soul out of it, so I like to put it to bed in eight hours or so. In three hours I want it to sound like a record with the basic sounds and feel. In six hours I should have all the balances and it should start to sound finished. After that the artist will come in for a listen.

Having the option to come back the next day is a great thing, though. When you come back fresh there's always a couple of obvious little things that you've overlooked. I find that towards the end of the day my ears get a little tired and I start to put a little too much top or echo on it.

*Where do you start your mix from?*
I have no system. I really work differently for every project and every different type of music. It's a matter of finding out what the center of the song is or what makes the song tick. Sometimes you build it around the rhythm section; sometimes you build it around the vocal.

Usually what I do is put up all the faders first and get a pretty flat balance and try to hear it like a song, then make determinations from there whether to touch up what I have or rip it down and start again from the bottom.

*If you're mixing a project, do you vary the sound from song to song or keep it all in the same sonic ballpark?*
The approach varies from song to song but I try to keep the same kind of reverbs and treatment for the drums. I try to keep some level of consistency but again, I'm also treating every song differently as well. I personally like records that take you to 10 or 12 different places.

*Do you add effects as you mix?*
I try to start out with a flat track, then find the tracks that are boring and add some personality to them.

*Do you have a standard effects setup?*
The only thing that I regularly do is to have like an AMS harmonizer on one stereo effects send with one side pitched up and the other side pitched down a little bit. On some projects I'm not using any reverbs at all, while on some projects I might be putting all my reverbs through Sansamps or some other kind of cheap stuff. I use a lot of things like Roland Space Echoes or stomp boxes. I feel that those things have a lot more personality than the high-end effect boxes sometimes.

*Don't you have a noise problem with them?*
Yeah [*laughs*], but I just make it work anyway. I'd rather have the personality with the noise than no personality at all. The cheap boxes have such character. There's a few boxes coming out now that have some color but a lot of the digital stuff is so bright that it just jumps out of the track too much. The new Sony box (the VP55) that I did some presets for is pretty good. I like it because it's kinda dark sounding but it finds its home in the track a lot better than the bright, clear digital stuff.

I love to have a real EMT plate or a real live chamber. For me, if I had just one good analog echo or reverb then I can make the whole record just fine as opposed to four or five digital ones.

*When you're using a real plate or chamber, do you go retro with some tape pre-delay?*
Usually I'll just use a DDL for that. Sometimes I'll use two sends into it; one that's straight into it and one that's delayed. I'll use the pre-delayed one for the vocal because of the space between the initial sound and the echo, which really separates the sound and makes it as big as possible.

*Do you have an approach to EQ?*
It's weird. I just use whatever it takes for the particular project. It depends on what's on tape, how well it was recorded, and how much work it needs. Bob Clearmountain is the genius for knowing what to touch and what not to touch and I think that's really the secret: what to fix and what to leave alone. I find that the more that I do this [mix], the less I actually EQ but I'm not afraid to put in a Pultec and whack it up to +10 if something needs it.

One thing that I use is a spectrum analyzer that I put across my stereo buss that lets me know when the bottom end is right or the esses are too sibilant. I know what a lot of records look like in the analyzer so I can tell when the overall frequency balance is right or might have some obvious little hole in it.

*Do you look for a specific curve or something that looks funny?*
I'm mainly looking at the balance of the octaves on the bottom end, like if there's too much 30 but not enough 50 or 80Hz. When you go into a lot of rooms, which is where the problem areas of the control room are, on certain consoles, depending on how the near-fields sit on them, there's a buildup of the upper lower midrange frequencies. So I look for those kinds of things.

*What's your approach to panning?*
The only thing I do is once I have my sounds and everything is sitting pretty well, I'll move the pans around a tiny bit. If I have something panned at 3:00 and it's sitting pretty well, I'll inch it a tiny sliver from where I had it just because I found it can make things clearer that way. When you start moving panning around it's almost like EQing something because of the way that it conflicts with other instruments. I find that if I nudge it, it might get out of the way of something or even glue it together.

*How do you deal with compression then?*
[*Laughs heartily*] Compression is like this drug that you can't get enough of. You squish things and it feels great and it sounds exciting but the next day you come back and it's like the morning after and you're saying, "Oh God, it's too much." So I've been trying to really back it off, especially with stereo buss compression.

What I will do a lot is buss my drums to another stereo compressor, usually a Joe Meek SC2, and blend that in just under the uncompressed signal. Sometimes what I'll do if everything sounds good but the bass and kick drum aren't locked together or big enough to glue the record together, I'll take the kick and bass and buss them together to a separate compressor, squish that a fair amount, and blend it back in. I'll add a little bottom end to that if the record still isn't big enough on the bottom. This helps fit the bass and kick lower on the record and gets it out of the way of the vocal.

*Do you use more delays than reverbs?*
Depends on the project. If it's a slick pop thing then I might use a lot of reverbs, but if it's a rock band then I might only use one reverb and maybe a half a dozen delays. I've tried really hard to wean myself from too many effects. I'll try to do different things like put only one instrument in the reverb, or put a reverb in mono and pan it in the same space as the instrument. I like the mono reverb thing because it doesn't wash out the track, especially if you EQ the return of the reverb so that it doesn't conflict frequency-wise with the instrument. I've done some fun stuff like compress the returns of the reverb so that they pump and breathe with the signal that's there. It gives the reverb a cool envelope that comes up after the dry signal and doesn't fight too much with it.

Before I approach a project, I know some basic things, like, if I want to make this record pretty dry or I've got an idea of what might work for this guy's voice. The one thing I will do is concentrate on giving the vocal the right character, so I'll fish through a bunch of different limiters to find out which is the right one or a bunch of different effects to try to find which one might complement his voice better, that kind of thing.

There was one record that I did where every time I put reverb on this guy's voice it just sounded like gratuitous reverb that didn't quite work, but without it, it was still missing something. The voice just wanted a little bit of sparkle. So I searched and searched and the only thing that sounded right for his voice was the old Ursa Major Space setting on the Cloud setting. I've done things where I've just put an old Eventide 949 on Random and run the input really hot and got this distorted kind of chorus thing that worked great. So you just gotta find that one thing that's like an article of clothing or something.

*What are you using for monitors these days?*

I've fallen in love with these Tannoy AMS10A's and I usually use those in conjunction with the NS10's. Every once in a while I'll go up on the big speakers if those are good. I might get my sounds at a pretty moderate to loud volume but when I'm getting balances it's always really soft. I listen in mono an awful lot and find it's great for balances. You can easily tell if something's fighting.

*Do you have any listening tricks that you like to use?*

I'll walk out of the control room and listen to it right outside the door. It's interesting to hear what it sounds like through the crack in the door. Things pop out. … Blasters are good things for sure as well.

*What format do you mix to?*

Mostly 1/2" analog because it adds a personality to the mix. I'll always do usually two DAT backups as well. Of late, I've used the Apogee 8000 converters, which sound really good. It's the first time that I've mixed through a digital device that sounds better coming out of the A/D than I mixed it.

*What gear do you generally bring with you?*

I've got tons of gear. Lots of limiters and Pultecs and API's and a lot of stomp boxes. For me, everything has to have a personality to give something a color so it pokes out of a track.

*How many mixes do you do?*

I'm really bad about that because I'll do a lot of options. I'll always do a vocal up in case someone at the record company complains that they can't hear a line. I'll always do a bass up or even a bass down as well. When I say up, I'm talking about a quarter or half dB because I find that if you get your balances good enough, that's the only amount of alteration you can make without throwing everything totally out of whack. A lot of times I'll do a number of other options like more guitar, more backgrounds, or whatever key element that someone might be worried about. And then sometimes if I'm not feeling like I got the overall thing right, I might do one more version that has a little tweak on that as well. Sometimes I'll add like a Massenberg EQ on the stereo buss and add a little 15k and maybe some 50 as well to give the record a little more of a finished master sound.

*How much time do you devote just to engineering these days?*
I work about 30 percent of the year just as an engineer and the other 70 as a producer. I won't take stuff on as a producer unless I truly believe in it. I feel like I have to understand the artist and be able to bring something to the project, whether it's vision, arrangements, sonics or all of the above. I've been offered a lot of great things but I haven't felt like I could add anything because it's great the way it is. There's no point in doing it unless I can take it to the next level.

# CHAPTER 15

# Lee DeCarlo

F rom his days as chief engineer at LA's Record Plant in the heady 70's, Lee DeCarlo has put his definitive stamp on hit records from Aerosmith to John Lennon's famous *Double Fantasy* to current releases by Rancid and Zak Wylde. If you ever wondered where those sounds came from, Lee takes us on a trip to find out.

*Before you mix, do you have the final product in mind?*
Before I even start to record something, I've already got the songs in mind, what order they're going to be in and what style they're going to be recorded in. A lot of times I'll start a record tight and finish it open. Tight, meaning not a lot of leakage, depending upon what the music is going to be doing. Because as the album grows, the music and the ambience grows along with it.

*Where do you start to build your mix from?*
The bass and drums. I'll put the bass and the drums up, and I'll get a rough sound on the drums real quick. I don't like to take a long time to do anything. I like to have it up and going. I just get the bass and drums so they just start to pump to where you can actually hear them breathing in and out with the tempo of the song. And as soon as I arrive at that, then I start adding the other stuff in.

*How would you do that?*
What I'll do is put the drums and bass in a limiter and just crush the hell out of it. Then I'll play with the release and the attack times until I can actually make that limiter pump in time with the music. So when the drummer hits the snare, it sucks down and you get a good crest on it, and when he lets go of the snare, the ambience of the bass and the drums suck and shoot back up again. You can actually hear a [*breathing sound*] going on that was never there before. But it was there, it's just that you're augmenting it by using that limiter.

*So are you using individual limiters on each track or is it just a pair of stereo limiters that you use?*

It's usually a mono limiter and it's usually something like an 1176 or a Summit or an Audio Design or something like that. It's whatever is handy at that particular point. Usually the best ones for doing this are the old Roger Meyer stereo limiters, but you'll never find one where both sides are working anyway so you do it in mono and bring it up the center.

*Do you have a method for setting levels?*

Yes I do. I'll have the drums around -5 with the snare drum constantly on the back-beat of the tune. From there I'll build everything around it.

A lot of people that really haven't been doing this that long think that what you do is just turn things up and add stuff on top of other stuff. So much of mixing is what you take away, either level-wise or frequency-wise. There are so many things that you have to eliminate in order to make it all sit together and work. Mark Twain once said, "Wagner's music is much better than it sounds." Wagner is a guy that wrote for cellos and French horns doing things in the same register, but it all worked. The only reason that it worked was he kept the other things out of their way. If you have an orchestra and everybody's playing in the same register, it's just going to get away on you. But if you leave holes, then you can fill up the spectrum.

*What would be your approach to using EQ?*

When I'm mixing I use a minimal amount, but when I'm recording I'm radical when I'm EQing. I do a lot on the recording side, but I'm just redefining what I'm doing on the mixing side.

*Do you use gates much?*

Sometimes. I may have a gate augmenting the snare, but it's in such a weird fashion. I always use the original sound of the snare drum, but I may have a gate on it that's so fast and has such a quick release that it sounds like somebody snapping their finger. I usually mix that in very low with just a ton of EQ on it, or use it just to send to an echo so that the snare drum doesn't have a lot of hi-hat or other things involved with it when it goes to the chamber.

*Are you adding effects as you go?*
Oh yeah, I would say 90 percent of the time I record with the effect on. Now, I may take the return of the effect and feed it to a track that's open. I will hold onto it as long as I can until I need a track if I don't want to go to two machines or something like that.

There are some things you obviously don't do, like vocals for instance. I always have people coming to me and asking me about, "How did you make John Lennon sound like that? What is the mojo filter that he puts on his voice?" There is no mojo filter. It's just John Lennon with a U87 and a 15ips delay. That's 133 milliseconds, or however many beats there are in the tune. I always put delays in [at] the tempos of the songs.

*Are you timing it to the snare drum?*
Usually I'll take an old UREI click generator and I'll find out what tempo the song is in. If you don't have a delay sheet, what you do is you start a stopwatch when the song is playing and you count to 25 beats. On the 25th beat you stop the stopwatch and you multiply that times 41.81 and you'll have how many beats per second there are. Then I'll set the click to that and I'll set up my echoes. I'll just get them so they pump. So when the click happens, you get the back beat or you get a 16th or a 32nd or a triplet or any sort of different returns for your echoes.

*Then you're delaying the chambers as well?*
No, I very seldom delay an echo chamber. A lot of guys do, but I don't. I much prefer to use the chamber just as it is, but I do use a lot of different chambers. I use somewhere around four or five different chambers on everything I do.

*How many delays would you be using?*
Probably three or four different delays, it all depends. I like little tricks with delays as well. I like to leave out the delay on maybe the last line of a phrase. Then everything has delay on it until the very last word of a sentence, or during an important statement.

*How long does it take you to do a mix?*

It all depends. These days it takes me a lot less time than it used to, but I was a lot more messed up in the old days than I am now. I can mix three songs in a day or I can mix one song in a day. But to be really comfortable, I like to take about a day to mix a song and then go away and come back and finish it the next day. If you can't do a song a day, then you've either got problems with the recording or problems with the band or problems with yourself.

*Do you use your own monitors?*

I have my own monitors but I tend to choose the studio for its monitors. I usually use Ocean Way (in Hollywood) because I love their monitors. Dollars to doughnuts, Allen Sides makes the best monitors in the world. There's nothing better.

*You're mixing on the big monitors then?*

All the time. Very seldom do I go downstairs. I have to feel my pants move when I'm mixing. I never sit down, I'm always dancing when I'm mixing. I have to feel the bass in my stomach and in my chest.

I do go downstairs (to the near-fields) to check out the relationship between the different instruments. You can't hear shit about the sound with them, but you can check out the relationship between them. But then you run into that trap of when you listen on a pair of NS10's, you get a lot more guitar than you thought you had. Then you bring the guitar down and when you listen to it on something else, it all goes away. So when I mix, I bring along an old pair of JBL 4311's and that's what I go down to.

*What level do you usually listen at?*

I like it loud. As a matter of fact, I'll start louder and work my way down. I'm always up there but it's not crushing. People don't come in and bleed from the ears, but I'm over 100.

*Do you have any special monitoring tricks?*

Everybody's got those. It depends on how comfortable you are with what you're doing. I find the more insecure you are about it, the more time you spend listening on different systems.

*How many versions of a mix do you usually do?*
I do a lot. I like to play around with it. I have always thought it would be a wonderful thing to mix your entire album in a day. And instead of doing one song a day for ten days, it would be a really great idea to mix the entire album ten times. Then go back and listen to which ones you like the best.

*What's your approach to panning?*
I have several different approaches. I like to pan stuff around a lot but I like to have the effects mono. And I like having things wide, but I don't like to have just a guitar on the right and the piano on the left. I've never been a big fan of that.

*Do you record stuff in stereo then?*
I record a lot of stuff in stereo. But I would rather do parts and have them mono. In other words, if you've got a guitar part playing in fours, I would have like one track of him playing two and four and the other track of him playing one and three. Then I could take that and move that around a little bit and then you get that sort of popcorn machine percolating effect. If you have the pump going underneath and that percolation on top, you can make the whole song come alive and fairy dust and sparks and magic will be coming off the whole thing, which is what it's all about. If you don't have the magic coming off of the tape, you might as well hang it up.

*So you bring the reverbs up in mono?*
Yeah, I always return them in mono. Just up the middle. See, the very nature of the echo (either a plate or live chamber) changes because they heat up during the day and cool off at night. What happens is if you have them panned left and right, you'll be sitting there and you get the mix of your life, and all of a sudden you realize all your echo on the lead vocal's only coming out of the left side. That's just a terrible disappointment.

*Do you have a special approach to treating lead instruments?*
Yeah, sure. Bass and drums are the heartbeat, just like a human body, but the face is what everybody sees. It's kind of like looking at a pretty girl. You see her face and her body, but what makes her run is what's inside. So the pretty girl puts makeup on and gets a boob job. In essence, I give singers and guitar players boob jobs.

*What are you trying to accomplish with effects? Are you trying to make everything bigger or to push things back in the mix?*
Bigger, wider and deeper. Everything has to be bigger always. Now, a lot of times I'll do stuff with no effects on it whatsoever, but I don't particularly like it. But, with effects you make a point about your music. Effects are makeup. It's cosmetic surgery. I can take a very great song by a very great band and mix it with no effects on it at all, and it'll sound good, and I can take the same song and mix it with effects and it'll sound fucking fantastic! That's what effects are for. It's just makeup.

*You're going for bigness rather than for depth, or both?*
I'm going for pump, always. If the song doesn't breathe, I fucked it up. The better the band, the easier the pump happens. Nothing happens if the band doesn't play it in the pocket to start with. There's not a damn thing I can do to fix it.

Everything has to breathe. Songs have a life and you have to develop that life within the song. Every single piece of music in the world breathes if it's played properly. A song is about something and the trick is to capture what it's about and make it live. That's why mixing's an art and not a technology.

What I do, and what the guys that are really good do, is we play a console. It's sort of like the infinite Mellotron, if you will. It's an actual instrument, and the guys that are good at what they do play it like an instrument.

# Benny Faccone

E ngineer Benny Faccone is unique in that he's a Canadian from Montreal, but 99 percent of the things that he works on are Spanish. From five Luis Miguel records to Grammy winner Ricky Martin to the Latin rock band Mana (also Grammy winners) to the Spanish re-mixes for Boys 2 Men, Tony Braxton and Sting, Benny's work is heard far and wide around the Latin world.

*What's the difference between doing a song in Spanish and one in English?*
First of all, the way they sing in Spanish is totally different than English. The syllables don't fit in the same way. If you notice with English music, it feels like the voice fits right into the music rhythmically. You can't do that with Spanish because it has different accents with harder esses. You have to treat it a different way on the mixing side by building the rhythm track around it. It's a different flavor with a different kind of emotion.

*Are there any other differences between doing an American record and a Latin one?*
Everything I do is treated like an American record. It may not be exactly what they want, but it's what I do. Even though the language may be Spanish, I try to treat it like I would an English record.

*Do you just do Latin Pop or do you do any traditional Salsa?*
As a matter of fact, I do everything. The Latin field is not very specific like the American market where you do one type of thing and that's all you do. In Latin music, you just do it all. I've done a couple of Mariachi records. There were a few records where they wanted some traditional salsa and the only way to get it was to go to Puerto Rico and do it there. I had to get some ideas of how to do it from some engineers down there since they have very specific placement for a lot of the instruments.

*And what is that exactly?*

They've got two or three different ways of doing it but the things that stay the same are the shaker and the bongos are always in the middle. Usually I do percussion as an overdub. I don't deal with a lot of big orchestras — the biggest is about eight or nine people with a basic rhythm section and a couple of percussion players.

*Do you find differences recording in other countries?*

If you go to Mexico, some of the studios are pretty nice but the maintenance is bad. I was recording a live concert once on two analog machines at 15ips where you start one and before that machine runs out, you start the next one to overlap. When I brought the tapes back here (LA) to mix I found out that the second machine was running at a different speed than the first machine, so you really have to be careful.

In Spain, they like everything new. It's really hard to find an analog machine since everything is digital. Whatever's the latest, that's what they want, but you can't find a piece of vintage gear at all. In South America, it's whatever you can get that works. Even in Miami, you have to fly a lot of gear in from New York.

*Is the use of effects different?*

In the Latin market, because you're not just working on one genre like Pop or Rock, everybody's different. One day I could be working with a producer who loves echo (reverb) and wants it real lush and the next day I could be working with a Rock band that doesn't want to hear any echo. One day I could be working with somebody that wants the vocals up front and loud and the next day with somebody that wants it as tucked in as possible. Rock has become very Americanized in the Latin market where they want it very dry, as is currently the trend.

*Do you have a philosophy or an approach to mixing?*

The only approach is to try to figure out the direction of the song, develop a groove and build it like a house. It's almost like a musician who picks up a guitar and tries to play. He may have the chart in front of him but soon he has to go beyond the notes in order to get creative. Same thing with mixing. It's not just a thing of setting levels any more, but more about trying to get the energy of the song across. Anybody can make the bass or the drums even out.

*How do you build your mix?*

It really is like building a house. You've got to get the foundation of bass and drums and then whatever the most important part of the song is, like the vocalist, and you've got to build around that. I put the bass up first, almost like the foundation part. Then the kick in combination with the bass to get the bottom. Because sometimes you can have a really thin kick by itself, but when you put the bass with it, it seems to have enough bottom because the bass has more bottom end. I build the drums on top of that. After I do the bass and drums, then I get the vocal up and then build everything from there. A lot of mixers just put the music up first, but as soon as you put the vocal up, the levels become totally different. After all the elements are in, I spend maybe a couple of hours just listening to the song like an average listener would and I keep making improvements.

*Do you have a method for setting levels?*

Yeah, I have a starting point. I usually start with the bass at about -5 and the kick at about -5. The combination of the two, if it's right, should hit about -3 or so. By the time the whole song gets put together and I've used the computer to adjust levels, I've trimmed everything back somewhat. The bass could be hitting -7 if I solo it after it's all done.

*Do you put the snare at about the same level as the kick?*

No. There, it's more a question of feel more than level. Because there's so many transients, it could be reading -10 and it could be still be too loud.

*What's your approach to EQ? Do you have certain frequencies that you always come back to on certain instruments?*

Yeah, as a starting point. But I'll do whatever it takes, depending on how it was recorded. For bass I use a combination of a low frequency, usually about 50Hz, with a limiter so it'll stay tight but still give it the big bottom. Add a little 7k if you want a bit of the string sound, and between 1.5 and 3k to give it some snap.

For the kick, I like to have bottom on that, too. I'll add a little at 100 and take some off at 400, depending on the sound. Sometimes I even take all the 400 out, which makes it very wide. Then add some point at 3 or 5k.

On the snare I give it some 10k on the top end for some snap. I've been putting 125Hz on the bottom of the snare to fill it out a little more.

For guitars, usually 1.5k gives it that present kind of sound. Pianos and keyboards vary so much that it all depends on how it feels in the track.

For vocals, it really depends if it's male or female. If they sing really low, I don't add as much bottom end. Usually I always take some off at about 20Hz to get rid of rumble. But anything on up, it really all depends on the singer. I might add a little bit in the 4–6k range in there.

*What's your approach to compression?*
Limit the heck out of everything [*laughs*]. I like to compress everything just to keep it smooth and controlled, not to get rid of the dynamics. But I don't like a compressor across the stereo buss because then it sounds like it's not breathing right to me. Even for Hard Rock, I don't like to do that. It's easier to do it individually.

Usually I use around a 4:1 ratio on pretty much everything I do. Sometimes on guitars I go to 8:1. On the kick and the snare I try not to hit it too hard because the snare really darkens up. It's more for control, to keep it consistent. On the bass, I hit that a little harder, just to push it up front a little more. Everything else for control more than sticking it right up in your face kind of a thing.

*Do you have any special effects tricks that you use? Any neat things you like to use all the time?*
I use a lot of the old PCM42's on guitars for a very short slap delay. It's mono but it sounds really big. I use something like 4, 8, 11 milliseconds, so it doesn't sound like a delay. Sometimes I use as much as 28ms on a power guitar. You stereo it out, it'll sound like two guitars on either side of the speakers.

*Is there a certain listening level that you always listen at?*
Yeah, I have my amps set pretty much at a certain level. It's a fairly modest level; not loud, not soft.

When I start the mix, I crank it a little bit on the big speakers to kinda get hyped a little bit and check out the bottom end. Then I'll slowly start listening softer and softer.

*How many versions do you do of a mix?*
Usually one with the vocal up .8 dB and another with the vocal down .4 dB. And if there's backgrounds, the same thing. I do not want to come back to remix. Once I'm done with a song I've heard it so much that I don't want to hear it ever again.

# CHAPTER 17

# Jerry Finn

J erry Finn is one of the industry's new breed of engineer/producers, brought up in the techniques of his successful predecessors yet willing to adapt those methods to fit the music and artists of today. From his mixing debut on Green Day's *Dookie* to their follow-up *Insomniac* to producing and mixing Rancid's *Out Come The Wolves* and *Life Won't Wait* to his work with the Presidents of the United States, to the Goo Goo Dolls and Beck, Jerry adds a distinctive edge loved by artists and listeners alike.

*I know you still do a lot of records that have relatively small budgets. How much of the money is spent on mixing in those cases?*
The majority. A lot of times when I get called in at the end to do a record like that, my mix budget ends up being at least twice as much as the budget for the rest of the album. My manager and I always try to work it out with bands that have smaller budgets though. I've done a lot of indie stuff with bands that were my friends for anywhere from free to half my rate just because I love the music.

*Do you usually have to work fast because of the budget?*
Not usually. I generally take about 10 to 12 days to mix a record. Some take less; some take more. *Dookie,* I think we did in nine days. *Insomniac* took 11 days.

I mixed Beck for a PBS show called *Sessions at West 54th.* We were supposed to only mix four songs in one day and it went so well that we ended up mixing seven songs in ten hours and it came out great. The stuff was recorded really well and his band had actually just gotten off a year and a half tour, so they just nailed it, so it didn't really require any fixing. And Beck is someone who really trusts his instincts so he doesn't sit there second guessing himself. We just went straight for what sounded right and just nailed it.

*Are you working 24-track analog?*

Yeah, almost all the time it's 24 analog. If I produce, I always try to keep it 24 analog. Rob Cavallo, who I mixed a bunch of stuff for, likes to do drums and then make a slave reel and work off of that, so those end up being 48-track. And actually, the last album I mixed for him ended up being 72 tracks — three analogs locked up.

*Before you start a mix, can you hear the final product in your head?*

Yeah, that's actually one of the requirements for me to feel comfortable going into a record. When I'm sent rough mixes, I really need to hear where I would take it in order to feel comfortable. Sometimes the band tells you what they want and the producer tells you what he wants and the A&R guy tells you what he wants and they're all completely different things. That can be a bit frightening because you end up being the punching bag for their arguments [*laughs*]. But I usually can hear the final mastered record from day one and then it's just trying to get that sound in my head to come out of the speakers.

*Where are you starting your mix from? Do you start from the kick drum, the overheads…?*

Just out of habit, I probably start at the far left of the console with the kick and start working my way across. Lately, I've tried to put the vocal in early in order to create the mix more around that. In a lot of the Punk Rock stuff you get the track slamming and then you just sort of drop the vocal on top. But for the poppier stuff I've found that approach doesn't work as well because the vocal really needs to sell the song. So I've been trying to discipline myself to put the vocal up early on, before I even have the bass and guitars in and kind of then carve those around the vocals.

One thing that I do with drums, though, is try to get the room mics in early on before I start adding reverbs and stuff like that to the snare. I try to push up the room mics and get the sound right on those, and I try to provide a lot of the drum ambience naturally, without going to digital boxes. Unfortunately, recording drums is sort of becoming a lost art. I mean, it is the hardest thing to record. And as engineers have gotten more and more dependent on samples and loops and drum machines, and with more recording being done in home studios, the thing that always suffers is the drums.

*Do you get a lot of stuff that's done in garages or homes?*
Not so much, but I do get stuff where the band thought that going to a good studio would be all they needed and they didn't really think about the engineer they hired. So I've seen some engineers that get in over their heads. I was actually a drummer myself when I played in bands so I tend to be real anal about the drum sounds. And I'm a complete phase junkie so if I think that the drums have been recorded poorly, it doesn't take much for me to criticize the drum recording [*laughs*].

*After you put the drums up, where do you usually go from there?*
I'll get the drums happening to where they have some ambience, then put the vocal up and get that to where that's sitting right. Then I'll start with the bass and make sure that the kick and the bass are occupying their own territory and not fighting each other. Sometimes, to my surprise, I've nailed it and it all falls together, and then other times when I get the guitars in there, they eat up a lot of the ambience on the drums. Most of the bands I work with tend to have several tracks of very distorted guitars and they want them all real loud, so then I have to go back to the drums and kind of adjust for that.

*How do you deal with that when you get a lot of real big crunchy guitars?*
When every guy in the band thinks he's the loudest, that's when I know I've nailed the mix. I've always tried to just make it so that you don't have to fight to hear anything. On certain parts of the song maybe I will bury something a little bit or push something a little louder for tension to kinda pull you into the next part, but overall I try to make it so you can hear everything all the time, and that generally comes through EQ. Like, I'll find the bite in the guitar and make sure that the snare isn't also occupying that same range. Then I'll make sure the low end on the guitars doesn't muddy up where the bass is sitting. And I also have to keep the kick and snare really punchy to kind of cut through all the wall of guitars by multing them off and hard compressing and gating them and sneaking them back in under everything.

*Do you find you use compression on a lot of things?*
Yeah. I'm a big compressor fan. I think that the sound of modern records today is compression. Audio purists talk about how crunchy compression and EQ is but if you listen to one of those Jazz or Blues records that are done by the audiophile labels, there's no way they could ever compete on modern radio even though they sound amazing. And unfortunately, all the phase shift and pumping and brightening that's imparted by EQ and compression is what modern records sound like. Every time I try to be a purist and go, "You know, I'm not gonna compress that," the band comes in and goes, "Why isn't that compressed?" So yeah, I compress the buss, although I'm very sparing on certain records. *Dookie* for Green Day had no compression on the buss at all and the Super Drag record that I produced and mixed last year didn't have any either. But if I think it's appropriate for the music, I'll get it on there.

*Are you compressing everything else individually as well?*
Lately what I've gotten into doing more of is multing it off, like I said. The kick and snare I'll put through maybe a 160 and very lightly compress it, maybe pulling down half to one dB. Then I'll mult them off and go through a new 160S and really compress those and sneak them up underneath so you're basically hearing the character of the drum you recorded rather than this bastardized version of it. Then I also send all of my dry drum tracks, not the rooms or overheads but the kick, snare and toms, through another compressor and sneak that in to give the kit an overall sound. Distorted guitars I don't compress as much because when you get a Marshall on 10, it's so compressed already that it doesn't really need it. But cleaner guitars or acoustic guitars, I'll compress. And I actually got into doing the vocals the same way I do the kick and snare; multing it off and compressing it real hard and sneaking that under the original vocal.

*When you say "real hard," how much do you mean?*
I would say 10 or 12dB and at a ratio anywhere from like 4:1 to 8:1. My compression technique is something I actually learned from Ed Cherney. He was telling me about compressing the stereo buss when I was assisting him, but I use the same technique on everything. I set the attack as slow as possible and the release as fast as possible so all the transients are getting through and the initial punch is still there, but it releases instantly when the signal drops below threshold. I think that's a lot of the sound of my mixes. It keeps things kinda popping the whole time. Also, you can compress things a little bit more and not have it be as audible.

*Do you have an approach to panning?*
Yeah, I tend to be a fan of panning things real wide. I think it started from when I was an assistant at Devonshire (North Hollywood). We had a (Neve) V3 over there that I worked on a lot, and when you engaged the pan knobs it changed the sound a little bit. So I tried to avoid using the pan knob by just assigning it to the left or right buss. Now I'll keep electric guitars, overheads, room mics and toms hard left and right, and hi-hat all the way to one side. There's not a lot of filling things in between.

The kind of bands I work with want to hit you in the head. For the most part, they're not really worried about having a Pink Floyd or Steely Dan–style mix where everything has its own spot. It's really supposed to hit you in the forehead, so the panning tends to be really extreme. Also, because radio tends to squash everything back up the middle, I've always found that panning it out like that makes it sound a little bit bigger on radio. If you take the stuff that's panned out wide and make it slightly louder than it should be in stereo, when you listen in mono it really comes together. I find that helps you avoid that all snare and vocal mix thing that you hear a lot of times and it keeps the guitars up there.

*Do you add effects as you go along or do you get a mix up and then add them?*
I'm pretty sparing on effects. Actually, over the last year and a half or two years, I've gradually tried to wean myself off of any digital effects. The last six or so things I mixed, the main vocal effect was a plate reverb and a tape machine or Space Echo for real tape slap.

*Are you delaying the send to the plate?*
Depending on the song. Sometimes it works, but with a lot of the music I do, the tempos are so fast that you don't really need to do much delaying because you can't really hear it. It's like the reverb needs to speak right away and then go away. I'm a big fan of the EMT250 on snare. That's probably been a standard since day one on my mixes. Electric guitars tend to stay dry and bass is always dry.

*What are you using for monitors?*
When I was an assistant, a lot of the engineers that I liked working with had Tannoy SRM10Bs. When I went independent, I searched high and low and finally found a pair. I carry those around with me wherever I go, as well as a Hafler Transnova amp, which gets frowned upon sometimes amongst the guys that are into the more hi-fi kind of thing. But I tried 20 amps and that just sounded the best.

*How much do you use NS10s or Auratones or the main studio monitors?*
I use the main studio monitors maybe 1 percent of the mixing time, if that. I know a lot of people say, "Well, I like to go to the bigs and listen to the low end and make sure that's in order," but the big monitors are so inconsistent from studio to studio that I can't trust them. Sometimes when the A&R guys come down, the band cranks it up to get them all excited and that's a fine use for $50,000 speakers.

I like to check my mixes in mono so I do actually use Auratones a lot. NS10s are sort of a necessary evil. Most producers and bands that I work with are used to them, so that's what they want to hear. But if I'm just listening for myself, I'll try to stay on the Tannoys.

*How loud do you listen?*
Extremely quiet. Like at conversation volume. Probably 85dB or so at the loudest.

*Do you usually mix by yourself or do you have people in the studio with you? Does it matter?*

It depends. When we did the *Dookie* record, the whole band was so excited by the whole process (they had never made a real record in a real studio before) that they were there the whole time with their elbows up on the console. On the flip side of that, sometimes the band and/or the producer will come in the first day and then I won't see them again. I was doing one record where the producer actually left the country and I didn't even know it. About four days into it, I said to the band when they came by to check the mixes, "Should we have the producer come back?" And they're like, "Oh, he's in England." So I guess he trusted me.

I like to keep the band involved and I always put their needs before my ego. I think a problem with a lot of mixers is the ego thing where when the band says, "You know that great sound you have? We want it to sound crappy." You have to take yourself out of it and go, "Well, their name's a lot bigger on the record than mine." Like on The Presidents' record, there were decisions they were making that I knew were gonna make the mixes sound weird sometimes, and I would explain that to them. Like, "Do you realize if you do that, it's gonna sound really strange on the radio?" or "It's gonna sound really sound strange when this chorus comes back in," and they'd be like, "Well, we don't care. That's what we want." So I'll do it and generally the band stays happy.

*What format do you usually mix to?*

1/2" analog. I like the old Ampex machines. They just kind of rock more than the Studers. Probably within the last year I've recorded, everything's at +3 at 30ips on BASF 900.

*How many mixes do you do? Vocal up, vocal down?*

If it's up to me, I'll do the main mix, a vocal up 1dB, a TV mix and an instrumental and that's it. I've worked with some producers that want to avoid conflict so they'll sit there and print mixes all day to please every guy in the band, but all you're doing is prolonging the argument. You end up with a nightmare at mastering as you edit between mixes, so I try to really just get it right the first time. Sometimes the A&R person will want vocal down, but...

*Vocal down?*

Every now and then someone will say, "Do a vocal down just to be safe," but I don't think it's ever been used [*laughs*]. I think some people just like to be covered and it's also probably people who are a little new to the business that think there might be some possible use for a vocal down, but there never is. The instrumental comes in handy sometimes for editing out cuss words and things like that.

*Where do you like to work usually?*

Conway (in Hollywood) is definitely my favorite studio. Before I went independent I was an assistant there for about four or five months. When I finally went independent, I was so scared because I had only done the Green Day record, but it was just blowing up so huge and I was getting so many calls that I had to pursue it. Being realistic about the music business, I thought I'd have a red hot career for six months and then be back assisting, so when I left I made them promise that when my career fell apart they'd hire me back as an assistant [*laughs*]. I still joke with Charlene, the studio manager, about that whenever I see her. "Are you still gonna hire me back when my career falls apart?"

# Jon Gass

abyface, Tony Rich, Mariah Carey, Usher, the *Waiting to Exhale* soundtrack; Mixer Jon Gass' credit list reads like a *Who's Who* of R&B greats. And with good reason. Gass' unsurpassed style and technique has elevated him to a most esteemed position among engineers, working with the best of the best on some of the most creative and demanding music around today.

*Do you have a philosophy about what you're trying to accomplish?*
Not really, I just go for it. I'm kind of a musical mixer. I grew up playing music and I'm not a real technical mixer at all. If something breaks, it's like, "Hey, it broke." [*laughs*]

I think I try to find the more natural tones of instruments and maybe boost that direction and make everything sound natural as long as it still fits together. I always think of it as a layer cake or something, so I just kind of layer the thing.

*Can you hear the final product in your head before you start? Do you know what you're going for?*
Actually, yeah, I can.

*What if you come in just to remix something?*
The last five or six years that's mainly what I've been doing. I know some people push up just the drums and work on them for awhile, but I start with everything on and I work on it like that. The reason is, in my opinion, the vocal is going to be there sooner or later anyway. You might as well know where it's sitting and what it's doing. All the instruments are going to be there sooner or later so you might as well just get used to it. And I think that's also what helps me see what I need to do within the first passage. That's when I start picturing. So it doesn't take me long.

*If you have something with a lot of tracks, let's say you're bouncing back between several vocal tracks, does it bother you that you can't hear everything?*

Yeah, if it's a real car wreck where you need to do a lot of mutes to make it make sense in the first place because there's stuff all over, then I'll actually go through and do a cut pass and get the cuts in it first so I can still kind of listen to everything. I'd rather do that so I can kind of get a grip of where it's supposed to go. If it's going to be one of those kind of songs I'll have the producer or whoever come down and help with the cuts first, and then send him off so I can get working on it. Most of the stuff I do, I really don't have producers or artists hanging out with me too much. It's really great for me. I think they like it, too. I guess I kind of have a style that most people that hire me think, "Okay, he's done this record and that record, so that's kind of what we want." If somebody helps me too much, I can only say, "Gee, you can pay somebody a lot less if you want to mix this. It's your budget." That allows me to be more creative.

And I'm scared to solo stuff a lot in front of the artists because I think individually the tracks that I mix almost have to sound bad. It really doesn't matter what it sounds like by itself though, because it has to work together. That's where some of the young producers blow it. They go through and solo tracks and make everything sound fat. Then they put it all together and have a big car wreck.

*How do you go about building your mix if you have everything up?*

I really start searching out the frequencies that are clashing or rubbing against each other, then I work back towards the drums. But I really try to keep the whole picture in there most of the time as opposed to isolating things too much.

*So you don't solo stuff much, then?*

Well, I do, but to solo something and EQ it is insane because it's not relative to anything — unless you're just going to do a mix with just that [*laughs*].

*What's your approach to EQ then? Do you just go through and look for things that are clashing?*

Basically, yeah. If there are two or three instruments that are clashing, that's probably where I get more into the solo if I need to hear the whole natural sound of the instrument. I'll try to go more that way with each instrument unless there's a couple that are really clashing, then I'll EQ more aggressively. Otherwise, I'm not scared to EQ quite a bit.

*You're doing mostly cuts? You're not adding?*
Yeah, especially on the SSL EQ. I'm definitely into cutting more than adding and I think that works best with the EQ on that board.

*How long do you think it takes you to do a mix?*
A day and a half is perfect. Two days is long.

*Do you have an approach to panning?*
Just balanced between left and right. That may differ a lot for different music I like to do. Playing so many years live, my left ear has been trashed by drums and hi-hat because the drummer was always on my left. But for some reason I always end up putting the hi-hat for the drums from the drummer's perspective as opposed to the audience perspective.

*What's your approach to compression?*
I'm pretty light on compression. Individual tracks, pretty light. Just really to add attack on acoustic guitars, electric guitars and stuff like that. Mostly on things I want to poke out of the mix.

*What do you mean by light?*
Unlike some of the New York guys that seem to use a lot of compression on a lot of things, I don't use that much. Same with the stereo buss. I barely touch it, with maybe a dB or two. Actually, I don't even use the SSL buss compressors any more.

*Just trying to even things out?*
Yeah. Just even things out, and I think too, if the stuff's EQed and layered right, you don't really need to do a ton of compression on the stereo buss. If the thing's laying right, at least with R&B, it just kind of sits there.

*When you're talking about layering, do you mean frequency-wise or level-wise?*
Frequency-wise. My ears have always been sensitive to frequency clashing, even back when I played in bands. I didn't know why, but frequencies onstage would drive me insane. Too much bottom maybe on the rhythm guitar amp clashing with the bass amp or something.

*When you're building your mix, do you look at the meteres and go by them? For example, the bass kick at –5, the kick at –5, etc. or strictly by feel?*

It's by feel. That's more part of the R&B thing too. Everything's kind of from feel to me. Sometimes the mix sounds great and somebody says, "The mix feels great, print it." As opposed to a Rock record where they might say, "More guitars," or something.

*Do you add your effects right from the beginning or do you wait until you have everything balanced out and then add them?*

As I go. I hardly ever use long halls or long reverbs. I use a lot of gear but it's usually for tight spaces. Sometimes in the mix it doesn't sound like I'm using anything but I might use 20 different reverb type boxes, maybe not set for reverb though, just to create more spaces. Though you may not hear it in the mix, you can feel it.

*How do you go about getting your sound? What determines what you're going to use?*

I don't have a formula. Whatever feels right. I usually have maybe 24 or 30 verbs and delays set up almost all the time. Not necessarily set to the same thing, but up. I think I have probably more outboard gear than anybody in the world. I like to use a lot different verbs. Instead of having 20 Yamaha reverbs, I'd rather have one or two Yamaha's and one or two Lexicons, because they seem to each have their own sound. The more different ones you use, the easier it is to separate the actual sounds.

*Before you start a mix then, do you have same effects set up all the time?*

Yeah. For instance, on the last song I did there was a Mini Moog type sound, and I had this kind of a short, tight room verb on it that set it back in the mix really nice. On the next song I didn't get a chance to change the setting, but I just happened to flip it onto the snare and it sounded great. So I didn't change it. It's the same effect from the last song, but it's on a completely different instrument. I have certain things set to what they do best, then I'll use them if I'm going to use that particular sound on a song. The next song I might not use it.

*Do you use mostly delays or reverbs or a combination?*

A combination. I do like the reverb programs with pre-delays and delays in them so that you can kind of customize them to the song and the tempo.

*So everything's timed to tempo, then, right?*
Depending on the song, yeah. Mainly the eighths, quarters, or sixteenths. But depending on the tune I'll add in triplets or whatever feels right.

The other thing I like to do with delays is to diffuse them. I'll put a delay through a bunch of stuff just to make it sound worse. We joke about this guy that mixed a long time ago, and he'd have his delay clearer and brighter and louder than the actual lead vocal. I think that's what kind of got me experimenting with ways to really tone it down.

Sometimes I put a delay through an SPX90 and won't even use the program. I just use it to clip the top and bottom end off and diffuse it off the lead vocal a little bit.

*When you're saying you use short spaces, are you trying to move stuff back or just put it in its own space?*
Yeah, put it in its own space. Sometimes it can be just a chorus, even a harmonizer with a really short delay time. What it comes down to is, I like short, dry sounds.

*How short?*
Like 25ms or less. I use a lot of 10, 12, and 15ms on things.

For *Waiting to Exhale,* for instance, a lot of that was really different for me because of the big string arrangements. That wasn't something that I'm used to doing, but I sure loved it.

*How did you approach the big strings? The traditional way, by putting a big hall on them?*
No, I kind of approached it differently. I didn't think that the stereo pairs were wide enough, so the first thing I did was spread them out about 10 milliseconds or so. Then I took the room tracks and kicked them back maybe 80 or 100 milliseconds, just to really make the room bigger. I was trying to create a bigger room on the room they already had before I started adding verb. And finally I just added a little bit of verb on the delayed room tracks. Once I created that, I thought it worked great. It's still kind of dryish, but it's gigantic. So it's really more of just the delays than verbs.

*Do you pan it opposite or just put it in back of the source sound?*
Usually stereo. In the R&B stuff you get a lot of stereo tracks
that really aren't stereo. That's one of the first things I do is
widen the thing out, even if it's only 3, 5 or 10 milliseconds, and
just get that stuff separated out so I can keep my center cleared
out. I don't really like that "everything mono" thing.

*With all the effects you're using, it sounds like there's a separate one for
each instrument.*
Absolutely. I very rarely use the same effect on more than one
thing.

*Do you use gates much?*
Lightly, especially on the SSL. I see a lot of the younger cats use
them too much along with too much compression.

*"Lightly" meaning the range is set where the level comes down a little bit
rather than off?*
Yeah, just a teeny bit. A little hiss is okay. Just the bad hiss is what
I'm trying to get out. But also, the SSL gates can sound a little
funny. I'll use a lot of outboard gates. Drawmers and stuff are
usually around for the hard work if I really need to do something more extreme.

*What will you use those on?*
Live drums and for triggers, stuff like that. I don't really use the
[Forat] F16 a lot for replacing drums, but what's great about it is
I can use the original drums and something from the triggers in
combination so that it's maybe not necessary to EQ the original
at all. I'll maybe add another kick that already has the frequency
I would've added to the original one. It seems like when you try
to add a lot of bottom or something to the original kick, it starts
to take away from the attack. So if you can leave that sitting the
way it is and add another bottom end kick to it, then you get the
best of both worlds.

*How about monitoring? Do you carry your own monitors with you?*
No I don't, but I really only work in about four different rooms.
The rooms I work in regularly have stock NS10s with extremely
high power on them and the mains are always TAD Augsburgers
tuned by Steve "Coco" Brandon.

And I mix really quiet on the big ones most of the time. That
seems strange but it's something that hit me about 15 years ago
when I went to my first mastering session and they were listening
quietly on the big ones and it sounded so good. And I was like,

"Wow! I could've made that sound better if I could've heard it this way."

*When you say quiet, you mean if you have a conversation you drown out what you're listening to?*
Just about, yeah. Like the SSL up on 1 (the Control Room Monitor level control) is what I mix on most of the time. It's really quiet but I can mix a very long time and not get fatigued. Sure, I do the NS10 thing, and then towards the end of the mix I'll go really loud on the NS10s and do some adjusting. And I'll go extremely loud on the big ones and do some more adjusting just to fine-tune. But I like it quiet on the big ones — but they have to be the right ones. I always said I could make a great sounding record on a cassette deck if I just had the right monitors.

*Do you ever use headphones?*
No.

*Do you have any listening tricks, like going down the hall or out in the car?*
I like to listen outside the room, but one of my favorite tricks is to turn on the vacuum cleaner and lay it up against the wall in the front of the room. Sounds a little strange, but I just kind of want to see if the mix is still cutting through at all. A blender would work, making margaritas or something [*laughs*].

*When you're doing a mix, how many versions of the same mix do you do?*
I'll do the main version, a lead vocal up, just the backgrounds up, and then the lead and backgrounds up. I hardly ever do a vocal down version. Then I'll just go through and pick some instruments that somebody might have questioned and put those up. It usually comes out to be 10 or 12 versions of each song, believe it or not.

*Covers your bases, though.*
If I don't do that, somebody always says, "It's too bad you didn't do one with this." But if I do that, it never happens. Even though they always pick the main version, I think people just feel better knowing that the alternate versions are printed.

*What format do you mix to?*
1/2" Studer at 30ips. I love the Studer machines the best.

*You have an interesting approach and it certainly does work. You try to make things bigger instead of washing them out.*

I think part of that is probably from my early recording days. I didn't really have any verbs so I had to use more of the ambience that was available. That started adding such a new twist as opposed to everything just miked so close and direct all the time. It adds such a great depth to everything.

*That must be the secret then.*

I'm sure that helps. But to me this business is about 95 percent luck, because if people don't call you and you don't have the right stuff to work on with the right gear, then it doesn't really matter. There's so many great, great engineers that are slow [work-wise]. It's really luck.

# Don Hahn

A lthough there's a lot of pretty good engineers around these days, not many have the ability to record a 45- to 100-piece orchestra with the ease of someone who's done it a thousand times. Don Hahn can, and that's because he actually has done it a thousand times. With an unbelievable list of credits that range from television series like *Star Trek (The Next Generation, Deep Space Nine* and *Voyager), Family Ties, Cheers* and *Columbo* to such legends as Count Basie, Barbara Streisand, Chet Atkins, Frank Sinatra, Herb Alpert, Woody Herman, Dionne Warwick and a host of others (actually ten pages more), Don has recorded the best of the best. Starting in New York City in 1959 and eventually becoming a VP at the famed A&R studios there and later at Hollywood's A&M studios, Don has seen it all and then some. He was kind enough to let me observe during a recent *Star Trek* session; then he shared some of his techniques and advice.

*How is your approach different from when you do something with a rhythm section?*
The approach is totally different because there's no rhythm section, so you shoot for a nice roomy orchestral sound and get as big a sound as you can get with the amount of musicians you have. You start with violins, then violas if you have them, then cellos, then basses. You get all that happening and then add woodwinds, French horns, trombones, trumpets and then percussion and synthesizers.

*What happens when you have a rhythm section?*
Then the rhythm section starts first. Any time I do a rhythm section, it's like building a building. That's your foundation. If you don't build a foundation, the building falls down. I like to shoot for a tight rhythm, not a big roomy rhythm section. I think that comes from all the big bands that I did; Woody Herman, Count Basie, Thad and Mel, Maynard Ferguson.

*Are you building from the drums or the bass first?*
The bass is always first. Everybody relates to the bass. I can remember doing records in New York and some of the producers would put me on and put paper over the meters. I told them I don't care; just let me get the bass and I'll balance the whole thing and it'll come out okay. The only time I can get screwed personally on any date with a rhythm section is if the bass player's late. There's nothing to relate to because everybody relates to the bass player. If he's not there, it doesn't work. Now orchestrally, like on *Star Trek,* the bass player can be late and it doesn't matter because I'm balancing all the other strings and then adding brass and the percussion last. So if the bass player's late, it doesn't matter. But on a record date with a rhythm section, it's the bass player and the drummer that's the foundation and the colors come from the synthesizer and the guitars.

*What's your approach to using effects?*
I'll use effects to enhance what I'm doing, but not to make like a bubble gum record. I don't do those kind of records any more.

A lot of the records that I do are, for lack of a better term, legit records. I've done a zillion Jazz dates. You can't put a room sound on a drummer on a Jazz date. It doesn't work; I've tried it many times. It ends up like a hot Pop record rhythm section and the music doesn't jive with it.

*I saw you using the EMT 250s the other day (at Paramount Studio M during a* Star Trek *scoring session). Wasn't the room big enough or weren't you getting the room sound that you liked?*
Well, you have to put some echo on it anyway so when you go to different studios and do the same show, it's got to sound basically the same every week. It doesn't matter what studio I go to, I still rent two 250s to make it sound consistent. Some of the studios have great plates but I don't have time to fool with them. When you're doing a live television show, there's no mix. You're mixing it as you're doing it.

*What would your approach be to adding reverb or echo?*
I mix emotionally until it feels good for me and hopefully it'll feel good for the producer and the composer and everybody else. I don't use a lot of effects, especially on the television shows. I do on records. Not a lot, but whatever I think is necessary if it's a little dull sounding. I can remember once with Earl Klugh, I had a popping rhythm section, but Earl plays an acoustic guitar and you can't put a lot of effects on it. So I had to tone down the rhythm section a little, otherwise it sounded like two different entities. You can't use a gimmick on an acoustic guitar like that, so it's sort of by feel. If the record doesn't make me bounce up and down, I'm doing something wrong.

*How about panning? When you're doing a* Star Trek *date, I'm curious how you're panning the various sections. Would it be the way the conductor's looking at everybody?*
No. I do that on movies. When I'm doing *Star Trek,* I do the high strings in stereo, the low strings in stereo, the synth in stereo, the brass and woodwinds in stereo, the percussion in mono, and whatever else mono. That's a stereo room and I pan it hard left and hard right.

*Do you ever worry about what it's going to sound like in mono?*
No, I check it and it changes a little bit, but it's not like a record because they add dialogue and sound effects. I used to worry about the studio and tape noise until I found out that every time they went into the spaceship on the show, there was a background hum in the ship. So now I get the least amount of noise that I can but I don't spend a lot of time fixing it because I'm not making a CD. You have to take all those variables into consideration because time is money.

*I notice you weren't doing much EQ or compression.*
I used a little bit. I think I had a little on the percussion, maybe a little top end on the cymbal and take some bottom end off the soft cymbal. But if you use the right microphones, hopefully you don't have to put that much EQ on anything.

*You're not doing much compression. Aren't you worried about somebody being out of control?*
Absolutely not. What're you going to compress?

*I assume on a record date it'll be a little different?*
Oh, yeah. You might get the French horns to jump right out at you. You might have to put a LA-2A on it and squash them just a little bit, but you shouldn't hear it.

*When you were doing the Sinatra dates, I assume it was all live.*
Yeah, I did Sinatra tracking dates in New York but Frank never showed up, so I've never personally recorded him. He called from his plane and said, "Just do the tracks, I'll overdub them in LA." And then on the *Duets* album, I did some extra vocals with Steve and Edie and I think Jimmy Buffet and Frank Jr. and I'm not sure who else. Now on the *Broadway* album I did maybe nine cuts, I'm not sure.

*Tell me about that. I'm curious if the vocalists are singing with the orchestra at the same time.*
Sure, that's the best way to make a record, especially with Sinatra or Tony Bennett or Streisand or any major artist. That's the way they're used to doing it and it's great. I mean, you really work your butt off, but you feel like you've accomplished something as opposed to sitting there all day and just overdubbing synth pads.

*What problems do you have in a situation like that?*
Headphones are the biggest problem in the studio. You never have enough separate cue systems to keep everybody happy.

*Are you worried about leakage?*
No, I try to get the least amount of leakage with as much room as I can. On Streisand, we put the bass player and the drummer in one section of the room with some gobos around, she was in her own booth, three other singers were in another booth and the whole rest of the studio was filled with great musicians.

*How has recording and mixing changed over the years?*
Well, just for some perspective, when I started there was no Fender bass and one track only, mono with no computers and no click tracks. Everybody played acoustic bass. There was no synthesizer. Bob Moog used to come up to the studio sometimes with his synthesizer (it was like 15 feet wide with big old telephone patch cords and tubes) and have us comment on his sounds.

I think some of the problems you have now is the younger guys don't go into the studio and listen. You must listen to what's going on in the studio. Don't just go into a control room, open faders and grab EQs. As an engineer you're supposed to make it sound in the control room like it sounds in the studio, only better. You must listen in the room and hear what it sounds like, especially on acoustic or orchestral dates, and not be afraid to ask composers. Your composers, and especially the musicians, are your best friends because whatever they do reflects on what you're doing. If they're not happy, you're not happy. Remember, the music comes first.

# Ken Hahn

T here are few people that know TV sound the way Ken Hahn does. From the beginning of the television post revolution, Hahn's New York-based Sync Sound has lead the way in television sound innovation and the industry's entry into the digital world. Along the way Ken has mixed everything from *PeeWee's Playhouse* to concerts by Billy Joel and Pearl Jam and a host of others while picking up a slew of awards in the process (four Emmys, a CAS award and 13 ITS Monitor awards).

*What's the difference between mixing for television and anything else?* Right away, the difference is that you have already got a certain restriction presented by the picture. In other words, the picture is only so long, so if you happen to get this great idea to do something that may change the length of what you're working on, it's probably not possible because the picture's already locked. If it's a half an hour show, that's how long it's going to be. It's something that most people coming from music can't get over. Like, "Wait a minute, what do you mean I can't fix this?" "Well, no, I'm sorry. The picture's locked." The reason why they can't go back is it's just too darn expensive.

Another major difference is that the deadlines in the TV world are a lot stricter. I always say, if it's in *TV Guide,* it's gonna be on the air. If they say the new so-and-so album's gonna be out the first week in April but it comes out the second week, then it's not as big a deal. But you never hear that *Barbara Walters Presents* will not be seen tonight because we didn't finish the mix. So there is a pressure on everybody to finish stuff, which in TV seems to be bad and getting worse.

*How long does it take you to do a typical mix?*
Well, it depends. We do a couple of series here where we get a couple of days to mix for a half-hour show, which ends up being about 20-something minutes of actual programming. So we need a day to do it and a day for people to see it and to do some changes. It ends up being about 16 to 20 hours, and that's for a show that's "together." You can do it in less, and you can certainly do it in more. News-style shows get less time and music shows get more, but I guess the answer is, never enough.

Video is now getting more like film in that they're doing more post-production on live shows. People now actually spend time previewing things, pulling sound effects, looping lines, and doing foley. In the last five or ten years, things are much more prepared by the time it comes to the mix. It used to be that you'd start at the beginning of the show and fly through it, just to make it digestible. But now it's gotten as sophisticated as film post-production, which can be very sophisticated.

*So essentially you have a lot of elements that you have to pull together.*
Yeah, it can be as big as a major film mix, 30, 40, 50, 100 tracks, depending on what's going on. The average viewer now doesn't know the difference between watching *Mission Impossible* on HBO and *Homicide*. They know one's a movie and they know one's a TV show, but when they're watching on a little TV they expect the same production value for either.

*With that number of elements, where do you start building your mix from?*
Most television and film is narrative in nature, whether there is a narration voice-over track that's telling the story or the dialogue is. Dialogue is premium, so most people start by making sure you can hear all the words. It's common practice here (Sync Sound) to do a pass mixing the dialogue, making sure that if nothing else played in the scene, the dialogue would still be seamless.

When you turn on the TV, the reality is that you set the level by the volume of the dialogue. You have to make sure all the words are in front and everything else is sort of window dressing. Music plays a huge role in it, too. What's been nice in the last few years is stereo television, which has only been around since MTV. Stereo music is a nice pad for things.

*Do you take advantage of stereo for anything else?*
Usually stereo ambiences like birds, winds, traffic. You can get
into a lot of trouble by panning effects too much. In film
mixing, at least you know that it's going to be played in a fairly
large room that has pretty good speakers. With TV, the listening
areas run the gamut from people laying in bed listening with
headsets on up to home theaters, so you have to err on the side
of safety which means put all the dialogue in the middle and
spread your music as much as you want left and right. But if you
start panning footsteps and all, it can really get weird because if
you're looking at a 15" TV while you mix and you pan footsteps
from left to right, then the panning will be all wrong if the
viewer happens to be watching on a 30" projection TV.

*Have you done much with surround?*
Music concerts, yes. But quite frequently, if it sounds good in
stereo, it's just going to sound better in surround. Inevitably,
once you kind of get the hang of it, you know what it's going to
sound like in surround anyway. The real battle has always been
to make it sound good on the lowest common denominator,
which is small speaker.

*What are you using for monitors?*
For a small reference speaker we use the staple of the industry,
the Auratone, but most of our stuff is mixed on bookshelf
speakers. We've used the KRKs a lot for the last five years. That's
pretty much what we've determined to be like an average stereo
speaker, yet it also relates to your average TV. We've done a
tremendous amount of listening to various kinds of TVs with
built-in speakers and found that the KRKs translate very well
from those speakers. That's what it's about, translating from big
speakers to small speakers.

*What level do you monitor at?*
I personally monitor about as low as most people would accept.
I tend to go that way because inevitably, if you get it sounding
good at a low level, it just sounds that much better at higher
levels. It sort of forces you to do a lot more manual gain riding
at low level because otherwise stuff just doesn't poke through.
I'm sort of doing my own form of manual compression and I've
found that usually works better than the other way around.

*Speaking of compression, how much do you use? Do you compress a lot of elements?*

I've done various things through the years. What's kind of cool about the Logic (AMS/Neve Logic 2 console) that we have, which has an all digital signal path, is it gives me multiple opportunities to control the gain. I do a little bit at almost each signal path, but I do it a number of times, some limiting, some compression, through so that it's pretty well controlled by the time it leaves here. Unfortunately, it's really frustrating to pop from ABC to HBO to ESPN and get radically different levels.

*Speaking of which, how much does everything change from what you hear in the studio once it finally hits air?*

It really depends on the network. It's incredible what sometimes happens to stuff on the air. It just flabbergasts me and my clients. We've delivered to anybody and everybody so we pretty much have an idea what you should do at our place before it gets to them so it will sound like you wanted it to sound like in the first place. You have to sort of put this curve on what you're monitoring so that you know that it'll sound fine on Viacom, for instance. I got a pretty good idea what HBO, NBC, etc. does to our stuff so you have to process the mix with that in mind.

*When you're remixing a live concert, since it's mostly music now, how are you approaching the mix? Where are you starting from?*

It's usually vocals again. I make sure that those are perfect so that it becomes an element that you can add things around. I always clean up the tracks as much as I can because inevitably you want to get rid of rumble and thumps and noises, creaks, mic hits, etc. Then I always start with bass and rhythm.

It sounds repetitive, but the vocal's where the story is. It's so integral to the music because that's where you're focused so it has to be as perfect as it can be. It can't be sibilant, tubby, too bright or too dull. It has to be properly processed so that it becomes another element that you have real complete control over. A guitar track for instance will probably be pretty consistent for the most part, but vocals inevitably are less controlled. The person may be on or off mic. They may be sibilant someplace, they may pop in other places. If you don't eliminate all those technical problems so that you can concentrate on the balance, you can really get bogged down.

It becomes even more critical in a production dialogue track where you've got, for instance, three people cut between different scenes and each sounds slightly different with slightly different room tone and different levels. Let's say you have a woman who speaks in a whisper with a guy who mumbles and another guy who yells. Well, if you don't level that out properly, you can't balance sound effects and music against it. I think that's the art of TV mixing. That's what makes the difference between people who really mix television and film for a living and anybody else. If you look at a film mix, there's three mixers and the dialogue mixer's considered the lead mixer.

*And you're cleaning those things up with the automation?*
Absolutely. Automated filters and just fader moves. That's one of the reasons why we got the console we did. It's completely dynamically automated, so you can roll in a high pass filter, zip it in and out, and the pop's gone. You can ride the EQ as you're trying to cover two people with a boom mic. If one is tubby and one is bright, you just literally ride the EQ through the scene until you get it right, so that it plays as close and consistent as possible. It wasn't that I was looking to get a digital console. I was looking to get a dynamically automated console and it happened that you got one with the other.

*Are you staying in the digital domain the whole time?*
Absolutely. I'll tell you, once you hear it this way, it's hard to go back to analog. What's different about television and film, as opposed to music mixing, is the number of generations that a particular track of audio may travel.

Let's say you recorded a location production soundtrack. It gets transferred to some medium and gets lined up with the picture. It now gets put into a workstation, then it probably goes back to tape of some kind. That individual track now gets pre-mixed to a dialogue track. So far we're talking like four generations already. Then it gets mixed into probably a final mix. That's five passes. Then it gets laid back to videotape. That's six passes and that's probably minimal for your average show. Most of them would go even more generations than that. With analog, there's just too many possibilities for phase errors, EQ problems, bias problems, noise reduction units being incompatible, especially noticeable when you mix for stereo. I mean it gets unbelievable. I've just found that the difference between analog and digital is just like night and day.

*Do you sweeten the audience much?*
Absolutely. I tend to make concerts sound as live as possible. I usually use a lot of the audience mics. I feel like the audience becomes another member of the band. The band is playing off of each other as much as they're playing off of the audience, so let's hear the audience.

*How do you deal with effects? Is it at the request of the act?*
I always try to become familiar with the material before I get to the mix so I know if there are any specific effects that are really important to the songs or to the artist. Also, a lot of people print an effects track that either you can use or get the idea from. But other than that, it's to taste. Luckily for me, people like my tastes.

# CHAPTER 21

# Andy Johns

A ndy Johns needs no introduction because we've been listening to the music that he's mixed for most of our lives. With credits like Led Zeppelin, Free, Traffic, Blind Faith, The Rolling Stones and most recently Van Halen (to name just a few), Andy has set a standard that most mixers are still trying to achieve.

*When you're building your mix, where do you start from?*
I don't build mixes, I just go, "Here it is" [*laughs heartily*]. Actually, I start with everything. Most of the people that listen to and tweak one instrument at a time get crap. You've just got to go through it with the whole thing up because every sound affects every other sound. Suppose you're modifying a 12-string acoustic guitar that's in the rhythm section. If you put it up by itself you might be tempted to put more bottom on it, but the more bottom you put on it, the more bottom it covers up on something else. The same with echo. If you have the drums playing by themselves, you'll hear the echo on them. You put the other instruments in and the echo's gone because the holes are covered up.

*Do you have a method for setting levels?*
That's all crap. That's rubbish. There was a famous engineer some years ago that said, "I can mix by just looking at the meters." He was obviously an upstart wanker. If you stare at meters long enough, which is what I did for the first 15 years of my career, you find they don't mean anything. It's what's in your soul. You hope that your ears are working with your soul along with your objectivity, but truly you can never be sure.

The only way that you can get a proper mix is if you have a hand in the arrangement because if you don't, people might play the wrong thing or play in the wrong place. How can you mix that? It's impossible.

The way that I really learned about music is through mixing because if the bass part is wrong, how can you hold up the bottom end? So you learn how to make the bass player play the right parts so you can actually mix. It's kinda backwards. I've been into other people's control rooms where you see them working on a horn part on its own. And they're playing with the DDLs and echoes and I'm thinking, "What are these people doing?" Because when you put the rest of the tracks up, it's totally different and they think that they can fix it by moving some faders up and down. When that happens, they're screwed. About the only thing that should move is the melody and the occasional other part here and there in support of the melody.

*Does the fact that you started on 4-track affect the way you work now?*
Yes, because I learned how to balance things properly to begin with. Nowadays, because you have this luxury of the computer and virtually as many tracks as you want, you don't think that way any more. But it was a great learning experience having to do it that way.

You know why *Sergeant Pepper's* sounds so good? You know why *Are You Experienced?* sounds so good, almost better than what we can do now? Because, when you were doing the 4-to-4 (bouncing down from one 4-track machine to another), you mixed as you went. There was a mix on two tracks of the second 4-track machine and you filled up the open tracks and did the same thing again. Listen to "We Love You." Listen to *Sergeant Pepper's*. Listen to *Hole In My Shoe* by Traffic. You mixed as you went along. Therefore, after you got the sounds that would fit with each other, all you had to do is adjust the melodies.

*What's your approach to using EQ?*
You don't get your sound out of a console, you get your sound from the room. You choose the right instruments and the right amplifiers for the track. If you have a guitar sound that's not working with the track properly, you don't use EQ to make it work. You choose another guitar and/or amplifier so it fits better in the track. It might take a day and it might take four or five different setups, but in the end you don't have to worry about EQ because you made the right acoustic choices while recording.

With drum sounds, even though where you put the mics is reasonably important, it's the way you make the drums sound in the room. The way you tweak them, that's where the sound comes from. The sounds come from the instrument and not from the mixer. On rare occasion, if you run into real trouble, maybe you can get away with using a bunch of EQ. But you can fiddle for days making something that was wrong in the first place just different.

*How about compression?*
I use compression because it's the only way that you can truly modify a sound because whatever the most predominant frequency is, the more you compress it the more predominant that frequency will be. Suppose the predominant frequencies are 1 to 3k. Put a compressor on it and the bottom end goes away, the top end disappears and you're left with "Ehhhhh" [*makes a nasal sound*]. So for me, compressors can modify the sound more than anything else. If it's a bass guitar you put the compressor before your EQ, because if you do it the other way around, you'll lose the top and mids when the compressor emphasizes the spot that you EQed. If you compress it first, then add bottom, then you're gonna hear it better.

*At what level do you listen?*
If I'm listening on small speakers, I've got to turn them up to where they're at the threshold of breaking up, but without any distortion, or I listen very quietly. If you turn it way down low, you can hear everything much better. If you turn it as far as it will go before the speakers freak out, then it pumps. In the middle I can't do it. It's just not Rock & Roll to me.

*Got any listening tricks?*
Obviously the idea is to make it work on all systems. You listen on the big speakers, the NS10s, out in the car, plus your own speakers, then you go home and listen again. This is a lot of work but it's the only way to go.

I tend to bring JBL 4310s, 12s, 13s and 12As and I put those out in the actual studio. But you know, I don't care how close you think you've got it that night, you take it home and play it back in the morning and every time, there are two or three things that you must fix. It's never happened to me where I've come home and said, "That's it." You hear it at home and you jump back down to the studio and sure enough, you hear what you hadn't noticed before on all the systems there as well. So every system you listen on, the more information you get. You can even turn up the little speaker in the Studer to hear if your mix will work in mono.

*Do you listen in mono much?*

No, but I'll tell you this: If you've got a fantastic stereo mix it will work in mono as well. For example, "Jumpin' Jack Flash" is a stereo mix released in mono. People don't listen in mono any more but that used to be the big test. It was harder to do and you had to be a bloody expert to make it work. In the old days we did mono mixes first then did a quick one for stereo. We'd spend eight hours on the mono mix and half an hour on the stereo.

*When do you add effects in the mix?*

I have some standard things that I do that more or less always work. I always need a great plate like an EMT 140 and a short 25 to 32ms delay just in back of the vocal. If it's kind of a mid-tempo tune then I'll use a longer delay which you don't hear because it's subliminal. It doesn't always have to be timed to the track; sometimes it can go in the hole so you can hear it. I've been talked out of putting reverb on electric guitars, but "Start Me Up" has a gorgeous EMT 140 plate on it. Most studios you go into don't even have one anymore.

*So you usually pre-delay the plate?*

Usually, but not always. In the old days, like on the Zeppelin stuff, you'll hear very long pre-delays on vocals. You know what that was? That was a 3M tape machine, which was originally designed to do video so it had about a 9-inch gap between the heads, as opposed to the 2-1/4" gap on a Studer or Ampex. Sometimes I'd even put it at 7-1/2ips. Another thing we used was the old Binson Echorec. Listen to "When the Levee Breaks." That was me putting two M160s on the second floor with no other microphones at all because I wanted to get John Bonham the way he actually sounded. And it worked! Page would say that he made me do it, but he was down at the pub. He did bring me his Binson Echorec for the track though.

*Do you prefer analog or digital?*

What I like is the sound that's coming into the mixer. I don't want it modified by some tape machine. I've always fought with analog. I've always fought with vinyl. With digital, the sound that's coming in, you get it back. It's much truer than any analog machine ever was. If you've got to smooth out your sound with some analog machine then you're in trouble to start with. With analog the noise factor is like a security blanket in that the hiss can cover up some weasely things.

*Which automation do you use, or do prefer to mix manually?*
They're all shitty because you're fighting a machine. I suppose the GML is the easiest but I still have to have somebody there with me to help. That's the part of the job that pisses me off. You've now got to be a bloody scientist. Sometimes it makes you too clever for your own good. If you just learn the tune then you're in tune with the tune. You let it flow through you. Now you might listen to it years later and say, "I think I missed that one." Or, you might go, "Fucking hell, I wish I was that guy again. That could not be any better. Who was that man?"

*The Mixing Engineer's Handbook*

# Kevin Killen

I n a prime example of how people interact today via high technology, I bumped into engineer/producer/ mixer extraordinare Kevin Killen via an Internet newsgroup. It seemed that a popular thread turned to how the bass sound on Peter Gabriel's "Sledgehammer" was recorded, and all manner of know-it-alls replied with the wildest of supposed methods and equipment, all of which were wrong. Eventually the real answer (Tony Levin's Musicman bass straight into the console with a little compression) emerged from the real voice of authority, Kevin Killen, who not only recorded and mixed Gabriel's seminal *So,* but also records by U2, Elvis Costello, Stevie Nicks, Bryan Ferry and Patty Smith to name just a few.

*Can you hear the finished product in your head?*
In certain instances I can. If I'm hired just to mix a project and I'm not intimately familiar with the material, I have just a general overview as to what I'd like it to sound like. As soon as I get in the studio, that's where I really start thinking about pushing or pulling a track one way or the other. For stuff that I've recorded, I usually have a pretty clear vision of what I want and actually try to start mixing as I'm recording. I like to work 24-track [rather than 48] so I try to make decisions based upon that. I'm always kind of mixing in advance.

*Where do you start your mix from?*
Usually the vocal. Maybe some of the rhythm section. I listen to what the strengths or weaknesses are and then build the track up around that. At some point maybe I'll just pop the vocal out and work on some of the rhythm stuff. I found that if I start with the vocal first I finish a lot more quickly rather than if I start from the ground up. If you're dealing with an artist who's a strong storyteller, that's going to be the main focus anyway.

*Do you have a method for setting levels?*

I've never subscribed to the point of view that there is a method. I just go with the flow. I had an experience about three years ago on a Stevie Nicks record with Glyn Johns, who's been making records since the 50's. We were mixing without automation and he would just push the faders up and within a minute or two he would have this great mix. Then he would just say that he didn't like it and pull it back down again and push it back up. I relearned that the great art of mixing is the fact that the track will gel almost by itself if it was well performed and reasonably well recorded. I find that the stuff that you really have to work a lot harder on is the stuff that has been isolated and really worked on. The tracks all end up sounding like disparate elements and you have to find a way to make them bleed together.

*What's your approach to EQ?*

I would imagine that I apply EQ based on my own hearing curves, whatever that is. I definitely hear a lot more high end than other people, maybe because my ears stick out and aren't pinned back flat to the head like other people. Because of that I tend not to over-exaggerate EQ. I try to get it sounding smooth. Most people mix in a much more aggressive fashion. I don't have individual instrument curves that I keep coming back to, because every bass drum is different, and every player is different, so I don't have particular settings or sounds that I go for, except to make it sound as musical and pleasurable as possible.

*Do you have an approach to panning?*

That's one of the things that I actually spend a lot of time on. I will get a balance that I like then I'll just try moving the panning around. I might spend a couple of hours experimenting because for me that is the kind of detail that can create a lot of space in a mix. I love to explore and create holes for instruments to sit in, but I'm not into gimmicks such as Spatializers to make the panning seem wider than the speakers.

*How about compression?*

When I can get it to work, sometimes I really like it. It's one of those things. I listen to other people's mixes and go, "That sounds amazing," but when I try it I can never get it to sound the same way. I tend to be quite modest on compression because my rationale is that you can always add more but you can never take it off. Since it will probably be applied at a later point during mastering and broadcast, I tend to err on the side of caution.

Since SSLs hit the marketplace I know what a temptation it is to set up the quad buss compressor even before you start your mix. I tried that for a while but I found out that I didn't like the way it sounded. What I came up with instead was almost like side-chain compression where you take a couple of groups on the console and you assign various instruments to them and use a couple of compressors across the buss and mix it in, almost as an effect, instead of using compressors across the inserts. You actually get a sense that there is some compression, yet you can ride the compression throughout the song so if there's a section where you really want to hear it, like in the chorus, you can ride the faders up.

*How about adding effects?*
If they're tracks that I haven't recorded, I get a quick balance with the vocal and the basic instrumentation to get a sense of the space around each instrument. If they've been recorded with a lot of ambience I'll shy away from it, but if the artist wants it to sound lush then I'll add some. Each situation is unique.

*Do you have a standard effects setup?*
I have some effects that I'll definitely go to but I won't necessarily have them set up beforehand. I really want to hear what's on tape before I start jumping in. What I normally request is a tape slap machine with Varispeed because it's still such a great sound.

*Do you bring your own monitors?*
I actually bring a set of English Proac Studio 100s and a Cello amplifier and my own cabling. Bob Ludwig at Gateway Mastering hooked me up with them and I've been using them almost exclusively for about three years. I find that when I take my stuff to mastering that it translates really well.

*Is there a difference between the way English engineers approach mixing
and the way American engineers do?*

I believe that the general trend in the UK is to enhance the
mixes with the appropriate effects, and there seems to be less
resistance to the notion of effects in general. In the U.S. it seems
a little more contrived; i.e., if you want to be a cool alternative
band you cannot use any reverb on the vocals, etc., etc. Personally I'm bored with that philosophy. Every recorded instrument
including the voice will have an ambience associated in the
room in which it was recorded. Therefore, I believe it's important to highlight the inherent musical quality of the performance. Of course an artist such as Elvis Costello likes his
material to be slightly less reverberant but I used more than he
ever knew because it was mixed appropriately. I try to show
reverence to the artist and the producer because when they
recorded the track they had a particular philosophy in mind.
I'm just a person to help them realize that vision.

# CHAPTER 23

# Bernie Kirsh

B ernie Kirsh has certainly made his mark as one of the top engineers in the world of Jazz. From virtually all of Chick Corea's records to working on Quincy Jones' "Back on the Block" (which won a Grammy for Best Engineering), Bernie's recordings have consistently maintained a level of excellence that few can match. Although technical know-how is all-important for an engineer these days, Bernie tells us that there are other, more human requirements involved in mixing as well.

*Can you hear the final result before you start?*
It depends on whether I've tracked it and I've been into it. If it's not something that I've tracked and overdubbed, then I'm discovering it as I'm mixing. But often, especially in the Jazz world, it's much more simple because I start out with wanting each individual instrument to have a pleasing quality. There's a preconceived notion I have of what that is. If you're talking about straight-ahead Jazz, there's a balance that's been accepted as part of the form. In that world, the cymbals are important, the position of the bass, piano, where the horns sit — all that kind of stuff has been listened to for decades. It's kind of a traditional form so that's somewhat predefined. If you move away or want to make a variation of that, then you're on your own. If it's something more in the electric vein, and something that I've worked on, then I'll come up with a notion of where I want it to go.

*How do you start to build a mix? Where do you start from?*
The first thing that I actually look for is the melody. After that, I'll go for the bottom of the mix.

*The bottom being the bass?*
The bass usually. I don't necessarily go for the drums first. Before I hit the rhythm, I usually try to get the melody and some sort of harmonic setup first, because I want that to be clear, and I'll often shape the rhythm to accommodate that. So that's the simplicity of it. If it's something that's more hard hitting, I'll

spend more time with the rhythm to get those guys pumping together.

*Do you have certain frequencies that you seem to come back to that need attention on certain instruments?*
Let's say for piano (which I've dealt with a lot), typically what happens is that in the analog domain it loses definition and openness if it's mixed some time after it's been recorded, so I'll usually boost in a couple of areas. First, up around 15k (sometimes that gets lowered down to 10 or 12, depending on the instrument), and maybe a little midrange at 3k or 5k. It depends on the instrument and setting, but that's pretty typical. I'll do the same thing usually with cymbals. I'll add between 12 and 15k on cymbals pretty typically. Those are the normal areas of EQ that I find that I'm constantly using.

*The frequencies that you adjust seem to be a little different than R&B or Heavy Metal.*
With this kind of music it's all about trying to go for more of a natural sound, for lack of better phrase. So, if there's going to be any hype at all it's going to be with the loudness button, where you get the larger bottom and accentuate the top. Normally, if you're going to add anything else to a piano, for instance, you're in the 500Hz range adding some warmth. But I find sometimes that when I finally get to mastering, the mastering engineer wants to take some of the warmth out for clarity purposes with just a little notch around 200 or 300. So, my tendency is to go for the warmth and then sometimes wind up taking some of that back out to achieve a little more definition or clarity later, if needed.

*Do you have an approach to panning?*
No, I normally keep things wide. Drums in stereo, piano open. I personally like a wide piano. I like it so that it feels like you're sitting at the instrument.

*You do it wide, left to right?*
Yeah, wide, left to right. I position everything as the player is seeing it rather than the audience. So the drums are from the drummer's perspective, piano's from the pianist's perspective, etc. — unless there's a leakage situation where I have to worry about the phase. If, for instance, the piano and the drums are in the same room, I have to make sure that the cymbal is appearing in the right place and isn't smearing because of the leakage into the piano.

*What projects are you the most proud of?*
What I'd like to do is delineate between the musical experience
and the audio experience because they're two different things.
There are albums that I did early on, which musically I enjoy
and I think at the time were sonically enjoyable. One was an
album called *The Leprechaun* by Chick Corea. That was 20 years
ago. The reason it was a lot of fun was because we just did live
recordings in those days. You had the horns and the strings, if
strings were there, and the rhythm section playing all at once.

It was the most fun because you became part of the creative
process. It's actually not a process at all; it's a different kind of
craft. A different type of musical creation. So I do enjoy that.

Although I didn't do the mixing, I did do some recording on
Quincy Jones' *Back on the Block* record a few years ago. That was
a great experience and it won the Best Engineering Grammy
Award. I actually learned a lot doing the record because it was
so different than the Jazz world, flying parts in and around, and
using all these techniques that people use that may not be used
in straight-ahead Jazz.

*There was such a cross pollination of different types of music on that.*
On that record, I was recording Rap, R&B and straight synth
parts. It was a playground in there. It was so much fun. Working
with a guy like Quincy was just a fantastic experience in itself
because this guy is a genius. He's a superb musician and he
knows how to work with creative people. He understands it and
he gets people to do what they do best.

*Is there a certain psychology that you use when recording?*
I wouldn't call it psychology, but it's in the realm of human
interaction. I've had people approach me and say, "Why don't
you tell people how you deal with others," meaning that they felt
good during this creative process, whereas in some instances
they haven't.

So what's the difference between how I was treating them as
opposed to how they've been treated by other engineers? I think
there are certain basic things that occur in that little microcosm
called a studio, which a lot of guys don't recognize. You're
getting into some basic human sensibilities that may not be
apparent as you look at it. For instance, you have artistic
creation going on. You have a guy who's come into the room,
who has done something that's very, very close to who he is. It's

not PR. It's not show. It's something that he holds very, very dear to himself. Now he's, for lack of better word, open and vulnerable and he's not being social.

So now you've got an engineer in the room whose attention isn't on that. Often you get engineers who, through various different bits of behavior, will invalidate the artist, evaluate for the artist, and not respect the frame of mind that the artist is in when wanting to make his musical statements. In other words, not looking at what the artist is doing at the moment. I think you'll find that the best engineers, the ones that the artists want to work with, have a notion that what the artist is doing is important and is something that needs to be treated with attention and respect. When I say that, I mean not to hold it up on a pedestal, but to understand that the action is something that's very close to the artist and not just a commodity.

For some reason, the creative process is different in the Jazz world. Guys are coming in, not necessarily to just lay down a rhythm track, but with the idea of making music. So I put a lot of attention on making the players happy with what they're hearing and make it comfortable for them. I don't work with a lot of engineers so I don't see it, but from the feedback I get, a lot of the younger guys don't recognize that element is really important. It seems like the job is really 10 percent technical. The rest of it's how you work with people and help them get what they want.

# George Massenburg

F rom designing the industry's most heralded audio tools to engineering classics by Little Feat, Earth, Wind and Fire and Linda Ronstadt (to name only a few), George Massenburg needs no introduction to anyone even remotely connected to the music or audio business.

*Can you hear the final mix in your head before you start?*
No. I generally look for a trace of feeling and I diddle things until I get a response. Whether it's EQing or changing arrangements, it's got to work as a feeling. And as such, I feel that what I do is significantly different from anybody else. I don't go into a studio to make money [*laughs*]. I go in to experiment.

*Is that a collective feeling or is it singular?*
Just about any successful piece of music is not something that can be performed by one person. It's almost always a collaboration. I can't think of anything that only one person has done in Pop music.

What I go after in mixing is a collaboration. Let me describe what I do with Linda (Ronstadt). I go after what I need and she tells me what she needs and then I try to steer a middle course between the two.

*When you begin to build your mix, where do you build it from?*
I always start Rock & Roll with drums, but very quickly I'll get a voice in there so that the instruments are crafted to work to the texture and the dynamics of the voice. I don't have any real rule. I actually can start just about anywhere.

*When you start with your drums, are you starting with the overheads first and building around that?*
Yeah, I generally will start with overheads.

*Room mics or overheads?*

Well, first and foremost I'm listening to the music, so I'll start with whatever gives me the best picture of what's going on in the room. I'll get a fast, overall mix and while I'm figuring out the tune, I'll start listening for problems or things to improve. Problems might range from a less-than-effective instrument amp or a mic placement to some big, funny boink somewhere that's sticking out. I like to tune things, line up overtones. I feel that equalizers are best used when used the least. I use them most to get rid of tones that are somehow not flattering. I'll most often use parametrics, sharp and subtractive, to look for the two or three biggest out-of-sorts characteristics. A snare drum for instance, has any number of boinks that I'll locate — I'll just have them ready to go on an equalizer — I may take them out or bring them up as I'm listening to the whole presentation, but I'll already know what and where they are.

*How about effects? Do you add effects as you go along or balance everything and then add them?*

I think of engineering as an art. I think that anything we do, we do for emotional reasons. So, the more we can keep what we do, the better. I start saving a mix on the very first monitor mix of the track. I'll save a snapshot and sometimes save the automation. What I'd really like to do is save everything, every reverb, and every delay. So if you think of what happens with a painter and a canvas, he picks up where he left off, and he may erase things or add things. If you X-ray a Rembrandt, there's all kinds of other ideas under the surface, but they start where they left off.

Well, we're not able to do that yet. We can't start where we left off because there's not the facility to do it yet. You get to a point in the track where everybody's really excited, but then you come back and it's completely different. "You know, it really sounded good the other night. Let's do that." But you can't get it back. Everything is subtly different and it throws you off. Have you ever tried to match a mix to a cassette that the artist brought in? It's impossible.

*You're pretty much staying with all the effects that you start with and just building on that then, right?*

Well, I may or may not stay with them. I have to see if they still count. Often delays and things that are musically related or effects the guys play to, especially like long reverbs, I'll print them. Delays, choruses, print them. Anything that's sonically significant to musical performance, I'll print.

*Do you have an approach to using effects?*

I don't have an approach. This is probably my biggest strength and my biggest weakness at the same time. I really try to invent everything from scratch every time I walk in. But yeah, I have basic things that I keep going to.

*When you're beginning to set up for a mix, are there certain boxes that you automatically put up?*

Yeah, there's probably eight different starting points and two or three of them will stay. The starting point is a plate, a delay, a [AMS] DMX, a [Roland] 3000 or 3500, a [TC Electronics] 5000 for a different kind of reverb, and a [Eventide] 2016. I'll use the 5000 for short stuff. I'll sometimes have a second 2016 for chorus. An [AMS] RMX is standard.

I'll have about eight delays set up. If I can send something into a delay, I'll do that because it takes up a lot less room. If I can make it sound like a reverb, I'll use it. I'll always go with the delay instead of a reverb if I can hide it.

*"Hide it" meaning time it to the track?*

Yeah, timing it so you don't really hear it as blatantly. You hear richness and warmth.

*And the timing is what? Eighth notes or dotted or triplets?*

No, it's musical. The timing will change. Often it's just by feel. I just put it up and try to get something that rocks.

*What's your approach to panning?*

I've got two or three different approaches and I'm always changing it. I used to be impressed by a drummer liking what I did, so I pretty much only got a drum perspective. But I've gone wide and I've gone narrow.

I've been working with Glyn Johns, and Glyn is a master of the accidental big airy drums, of course with Led Zeppelin. It's a great story. I was having dinner with Glyn and Doug Sax one night, and he was telling us about the first Led Zeppelin record and how they set up the drums in mono. They had one 67 right over the snare but they always needed a little bit more floor tom so he stuck a mic at elbow level, kind of off by the floor tom, pointing into the snare. After he finished the track he grabbed the mic and put it on the guitar and panned it. When he put it back on drums he forgot to pan it back. "Oh, that sounds great. I wonder what happens if I take the overhead and pan it right?" And Doug Sax and I looked at each other and said, "You got stereo drum miking by accident?" And in that case he became well known for that big airy Led Zeppelin and The Who sound. It was a different sound than what was being done in New York, which was almost all mono, or California, which was a spaced pairs kind of thing. The earliest stereo that I knew didn't even include stereo drums.

*Is there a general mixing approach that you have?*
I want to hear something authentic. I want to hear an authentic room or an authentic performance. I want to hear authentic instruments. It's not necessarily a sophisticated or elegant thing. It's just authentic. In stereo I try to paint a picture that makes sense, that your brain doesn't say, "Hey, what are you trying to put across on me?"

*How are you applying compression during the mix?*
The big difference between engineers today is the use of the compressor. At one time or another I tried to compress everything because I was building a compressor and I wanted to see how it did on every instrument. I'm a little off on compression now because there are so many people that overuse it. Everything is squeezed to death. As a result I'm backing off. When anybody goes that far out, I'll go the opposite way as hard as I can. But generally I will pretty much always have an option to compress the mix. I'll use my EQ, my compressor, then my converter and an M5000 to do three-band. Then I can dial it up from extremely subtle to pressed ham under glass.

I'll always compress vocals. I may recompress vocals again during the mix. I'll almost always have a bunch of compressors if I have to bring an element or a group of elements together like a background vocal, level them, then drop them into a pocket. Then I'll do some extreme stuff like compressing a room, then gating it. Maybe I'll compress a drum room and then gate it with the snare drum to get a real rectangular reverb. I do that a lot. Maybe I'll add reverb to a guitar and then gate the result of that. I do that some. Boy, I wish I could give you a rule.

*What are you trying to accomplish?*
Trying to get a thrill [*laughs*]. I'm almost always trying to get, as Lowell George used to call it, "decibel excursion," which is a bullshit term, but I love it. I try to make an instrument denser or give it some weight. Half of it's reverb or ambience, and the other half is bringing that ambience right up in your face, which is compression.

*How about monitoring? What's your typical monitoring setup?*
I started using Tannoys in '79 and '80. They do one thing. I love Genelec 1032s, they do another thing. We used KRKs for the Journey record because they were little rockin' monitors. Light them up and it's a completely different mix than what lights up Tannoys. And Yamahas, except that whatever lights those up makes for boring mixes. For rocking monitors, I'm just looking for something with that impossible to describe, lively factor. I don't know what it is.

I monitor on a lot of different things. I might go up to the wall monitors to try to hear subsonics. I'll go to Yamaha's to hear what the idiots at the record companies are listening to. Tannoys for fun. KRKs for fun. Earphones.

*You listen on headphones?*
I listen on headphones because you can hear if you're making a mistake.

*I always put up a set of headphones myself, and when I don't do it, I'm sorry.*
Yeah. You know who taught me that was Jimmy Johnson. He would always find that snap in bar 30 of the sax solo and you'd listen to it and sure enough, a tiny little snap to get rid of. And for the kind of music he was doing, that was appropriate.

*What levels do you usually monitor at?*
Everything. I'll monitor way loud to see what rocks. I'll monitor at a nominal level to get sounds together. Then I'll monitor about 5dB over background noise to hear all the elements into focus. If a mix works at 30dB SPL, 25dB SPL, it'll almost always work a lot louder.

*What are you listening for down that low?*
That the instruments work together. That you don't lose anything. If you can hear everything at that low a level, then when you turn it up you'll have a very even balance. That's the way to get everything in the same plane, by listening extremely low.

*Do you have any playback tricks? Do you go outside in the lounge and listen through the walls sometimes?*
All the time. I'm a big one for hallway. I hate cars. Through the control room doors is always an important thing for me, because I almost never do loud playbacks. I like listening around the corner and on a blaster.

*How many versions of a mix do you normally do?*
Well, I prefer to have one version of a mix. It's the same theory
as a horse race. You never bet against yourself. You never bet on
two horses to win. I believe in one mix and I believe either it's
right or it's not right. I will walk out of the control room with
only one mix. It's possible these days to do up and down in half
dB steps, but I don't really do it. At around that point, it's
important for you to let go of a mix. It doesn't belong to you
any more.

*Do you go back often and do any touch-ups or remixes?*
Yeah, all the time. But I think usually we go from scratch again.
Some of the best mixes I've done have been the fourth or fifth
or sixth passes. I remember on "Shinin' Star" we kept going
back in the studio and tracking and I think the more we went
back in, the more we found what didn't work.

*It seems like it's so much easier to refine things as you go back like that.*
Oh yeah, because you know what your priorities are. You know
what doesn't work, because the first couple times you go in,
you're trying exotic EQing and delays. You go back in and it
doesn't make any difference except for the stuff that's right. The
thing that makes a difference is vocals. I've spent more time
than anything else trying to find how to do vocals and how they
tell the story.

*What do you look for in a studio? Is there anything special that you
have to have?*
Good food! [*laughs*] It has to have that vibe. You have to take
the studio seriously. You have to walk into a studio knowing that
great music has been made there. Yeah, I need that in a studio
because then I rise to the challenge. If you go in and record
strings at Abbey Road Studio 1 and they sound bad, you know
that you messed up [*laughs*]. When Linda Ronstadt opened
Skywalker with her record, we (including Peter Asher, the
producer) intended to make a fairly large record. We used every
aspect of that big room. It won two Grammys and sold three
million copies right away and had two substantial hits. Well, the
next guys that came into Skywalker could say that they didn't
like our record or our music if they wished, but they certainly
couldn't say that the room didn't work. So I look for that vibe. I
look for whether really successful music — preferably music that
I love — has been made there. If I went to Rudy Van Gelder's to
record Jazz, I would be really motivated to get it right.

*If he let you in. That's the problem.*
He wouldn't. We know that. Herbie Hancock said Rudy Van Gelder yelled at him once and it's the first and last time he ever tried to lift the lid on the piano. Rudy came shooting out, "Don't touch!"

What you bring to the table in the control room seems to come through. I've been gifted to work with great musicians and any of the sounds that we get, any of the sounds that any of the really good cats get, it's because of great musicians.

# David Pensado

O f all the genres of music, mixing R&B may be the toughest, thanks to the almost constant change in the state-of-the-art and the penchant by the participants to experiment with new sounds. Mixer David Pensado, with projects by Bel Biv Devoe, Coolio, Take 6, Brian McKnight, Diana Ross, Tony Toni Tone, Atlantic Starr and many more, has consistently supplied mixes that have not only filled the airwaves, but ranked among the most artful as well.

*What's harder to mix, an R&B or a Rock track?*
I mix both and R&B is infinitely harder to mix than Rock. Think of it this way. Let's say you're painting a portrait. Rock is like having the person you're painting sitting in front of you and you look at them and paint, you look at them and paint. So you have a reference. In R&B, there is no reference. It's like trying to do a portrait from memory, but because you don't have the person there, you can paint something that transcends what he is. You can make him prettier, you can make him uglier, or you can make him abstract if you want. Doing R&B you've got less limitation and a lot more freedom. We don't have to have the snare drum sound a particular way. It can sound like anything from an 808 to a hand clap to a little spitty sound to a Rock sound. But you put certain snare sounds in a Rock song and it's just not a Rock song anymore.

*Do you hear the finished product in your head before you start?*
Yeah. I really can. I might not have 100 percent of the final product in my mind when I start, but I pretty much have it outlined. Then as I start filling in the outline, sometimes things change a little bit. Every once in awhile, maybe out of two or three hundred, I might just pull the faders down and say, "I don't like any of this" and start again from scratch.

*What's your approach to using EQ?*
Well, I think of EQ as an effect much the same way you would add chorus or reverb to a particular instrument or vocal. Like, I might have a vocal where I think it's really EQed nicely and then

I'll add a little more 3k just to get it to bite a little more. Then it just makes me feel like the singer was trying harder and it brings out a little bit of passion in his or her voice. So I tend to be most effective when I do the standard equalizing, then take it to the next level, thinking of it as an effect. Some of my favorites for this are the NTI EQ3, API 550 and 560s, the old "Motown EQs" at Larrabee and the Avalons.

*Are there certain frequencies that you keep on coming back to?*
I notice that in a broad sense there are. In other words, the frequencies from say 120 down to 20 cycles, I'm always having to add. It seems like the frequencies from say 10k up, I'm always having to add those. A lot of the music I do has samples in it and that gives the producer the luxury of pretty much getting the sound he wanted from the start. In the old days you always pulled out a little 400 on the kick drum. You always added a little 3 and 6 to the toms. That just doesn't happen as much any more because when I get the tape, even with live bands, the producer's already triggered the sound he wanted off the live performance and the drums are closer. It frees me up because now I have the luxury to really get inside the tracks within the time frame I'm given, whereas before I would have to spend that time just getting it up to a certain level. Now, in most of the stuff you're given, it's really starting out a lot better than it started ten or 15 years ago.

*How about panning?*
I think that there's three sacred territories in a mix that if you put something there, you've got to have an incredibly good reason. That's extreme left, center and extreme right. I've noticed that some mixers will get stereo tracks from synthesizers and effects and they just instinctively pan them hard left and hard right. What they end up with is these big train wrecks out on the ends of the stereo spectrum. Then they pan their kick, snare, bass and vocals center and you've got all this stuff stacked on top of each other. If it were a visual, you wouldn't be able to see the things behind the things in front. So what I do is take a stereo synthesizer track and I'll just toss one aside because I don't need it. I'll create my own stereo by either adding a delay or a chorus or a pre-delayed reverb or something like that to give it a stereo image. I'll pan maybe the dry signal to 10:00 and then I'll pan the effects just inside the extreme left side. I would never put it hard left because then there's too many things on top of it. I would pan it at 9:00, and then pan the dry signal to say 10:30, something like that.

*Do you use a lot of compression?*
There again, I look at compression as having two functions. One as an effect and when you want to keep a particular sound right up front in your face in the mix. I use quite an array of compressors because each one seems to give a little different characteristic as a result.

*Do you compress individually or on the stereo buss, or both?*
Well, I do both. There's a trick that some of my favorite New York mixers do to get the drums really fat and in your face. They would feed a couple of busses to a compressor and EQ the compressor output, then they feed kicks and snares and things like that to that compressor and just squeeze the heck out of the sound source. It literally is thought of and treated just as if it were a reverb or a chorus. In other words, just treat it as an effect that's mixed in with the original signal. More often than not, you're compressing the individual sounds as well.

I recently read an interview with a well known engineer where he was praising a particular compressor for its ability to take the dynamics out of a drum performance because the drummer would get happy on the first downbeat of every chorus and play a little louder. I thought, "I spent my whole career trying to add those dynamics and trying to make the drummer sound like he got happy going into the chorus." I very rarely use a compressor to even out dynamics. Dynamics are something that I just can't get enough of. The compressors I like the most tend to be the ones that actually help me get dynamics. That might be a contradictory statement, but if you're careful with the attack and release times, you can actually get a compressor to help you with it.

*Most mixers I've talked to don't think of their compressors that way. What do you use in particular to do that?*
Well, for kick and snare I'll use a 160X and I'll set the ratio at two or three to one, depending on how much transient response is already there. The Over Easy button will not be in. It ends up where I'm knocking off sometimes 20dB and no less than 15. There's a point at which you get an amazing attack in the range from about 400 to 3k. Then I'll take the output of that compressor and I'll feed it to a Pultec or a Lang or an API 550 and I'll add back sometimes 15, 20dB of 100–200Hz and a little 10k, and sometimes even 3–5k. Then I'll get my original sound where I like it and I'll add in that compressed sound. Man, it just puts the drum right in your face and makes it huge and fat. Basically what I'm doing is trying to take the frequencies

that I want and add those back into the original sound in such a way that I can't do with EQ.

A lot of times what I'll do is put the effects only on the compressed sound. In other words, an effect I use a lot would be "Locker Room" or "Tile Room" on a PCM70 and I'll add that effect only to the compressed sound. As a result, the reverb actually has a snap and aggressiveness to it. Every once in a while I'll make it stereo where I'll take two 160s and I'll set them up identically, but on the insert of one I'll put like anywhere from a 9 to 15 millisecond delay so the tight compressed sound is out on the edges of my stereo spectrum, but the original sound's in the center. That creates an incredibly nice image, particularly for ballads and slow tunes where you have a lot of space between the downbeats. That setup works great for snares, kicks and hi-hat. Every once in awhile it'll make a guitar come alive too.

So what you're doing is you're controlling the dynamics but you're actually increasing the dynamics. It's the strangest thing because psychoacoustically, it's not getting louder but your mind is thinking it is. On the radio, it just jumps out of the speakers.

*Do you have a philosophy about adding effects?*
The way I think of it is the pan knob places you left to right while the effects tend to you place you front to rear. That's a general statement, but it's a good starting point. In other words, if you want the singer to sound like she's standing behind the snare drum, leave the snare drum dry and wet down the singer and it'll sound like the singer is standing that far behind the snare drum. If you want the singer in front of the snare drum, leave him dry and wet down the snare drum.

That said, I like a vocal mostly dry, but then it usually doesn't sound big enough. You want the vocalist to sound like they're really powerful and dynamic and just giving it everything, so I'll put an 8th note delay on the vocal but subtract a 16th, a 32nd or 64th note value from that 8th note. What it does is gives a movement to the delay and makes the singer have an urgency that's kind of neat. So put the eighth minus one 64th on the left side, and put the 8th note straight on the right side. You can experiment with putting the pitch up a little bit on one side and down on another. If your singer's a little pitchy, that usually makes them sound a little more in tune. Sometimes putting the 8th note triplet on one side and the 8th note straight on the other (if you've got any kind of swing elements on the track) will make the vocal big, but it doesn't make the singer sound like he's taken a step back.

Another thing I like to do is to take the output of my effects and run them straight into another effect. I'll take an exciter and just dump the output straight to a chorus so it's only chorusing the high frequencies. I think that's more pleasing than having low notes chorusing all over the place. Another thing I'll do is set up an SPX1000 or SPX90 both on chorus. I'll put one where I'll pan it hard left and then I'll pan the right return at 2:00. Then I'll take another SPX90 and I'll pan it hard right, and then the left return from that one I'll pan at 10:00, so now the left and rights are kind of overlapping. On one I'll have the chorus depth just a little less than the other and I'll have the other modulating a third faster. When you add a vocal to that, you get this real nice spectrum that just widens because you're sending the both of them an equal amount but yet one of them is chorusing deeper and slower than the other one. If that's not wide enough for you, add a delay in front of both of them that's different on each side and then add that to your background vocals. They don't take any steps back in the mix, but they just get fat.

A lot of times I'll take two PCM70s and instead of running them stereo, I'll run them mono in and mono out and pan one just inside the left and one just inside the right. I'll use the same program on both but I'll slightly alter the values. Even if you don't use two PCM70s, just return the darn thing mono and you'll be surprised at how much better it sounds.

*What monitors do you like to work on?*
For the main monitors I like the Augsburgers with TAD components and for small monitors I use NS10s with the old tweeters. I also use Auratones, but in an odd way. A lot of times I'll start EQing my kick drum on the Auratones, which is kinda strange because you're adding a lot of frequencies that you can't hear, but you can see your meters going up. It forces you to EQ higher because if you're sitting there listening to dual 15" speakers and you're adding 20dB of 40Hz, you think you're killing the world. You go to the Auratones, you can't hear any of it so it's useless. So a lot of times I'll use the Auratones to EQ my extreme low and top end. You think you're adding high end when you're adding 10, 12, 14k but really what you need to be adding is 5k, and you'll put it on the Auratones and then it'll make it more honest and work within what is the real range.

Then I'll go up to the big ones and I'll watch my meters and make sure that I'm not getting too crazy, and then I'll add the super low stuff and the super high stuff.

*What level do you usually listen at?*
I usually listen to NS10s kind of medium and Auratones I listen at the same volume you would listen to TV. I found that on the NS10s, in order for them to really work, it's best to have them stay at one level for most of the mix. Then near the end of the mix, check your levels and your EQ with the NS10s about 20 percent lower and again about 20 percent higher and you'll make adjustments that you'll really be pleased with when you hear it on the radio. The big speakers I use mostly to show off for clients and to just have fun. I like to turn it up and if my body is vibrating properly, then I'm happy with the low end. A lot of engineers use them to hype the client, but I also use them to hype myself! If I'm cranking and I'm not getting excited, then I just keep on working.

# Ed Seay

G etting his start in Atlanta in the 70's by engineering and producing hits for Paul Davis, Peabo Byson and Melissa Manchester, Ed Seay has since become one of the most respected engineers in Nashville since moving there in 1984. With hit-making clients such as Pam Tillis, Highway 101, Collin Raye, Martina McBride, Ricky Skaggs and a host of others, Ed has led the charge in changing the recording approach in Nashville. As well as his insightful observations about mixing, Ed describes the evolution of the sound of Country music to what it is today.

*Do you hear the final product in your head before you begin to mix?*
To some extent I can. I think one of the things that helps me as a mixer, and one thing that helps all of the ones that have made a mark, is what I call "having the vision." I always try to have a vision of the mix when I start. Rather than just randomly pushing up faders and saying, "Well, a little of this EQ or effect might be nice," I like to have a vision as far as where we're going and what's the perspective. Definitely, I try to grasp that early on.

*Is there a difference between mixing Country music and what you did before?*
Country music is definitely lyric driven. One of the mistakes that some people make when they try to work on the stuff is they tend to downplay the lyric or downplay the lead vocal. And at first, I think some people begrudgingly push up the lead vocal and just say, "Listen to how loud it is." But there's actually an appreciation for having a really great vocal out there with a great emotion selling lyric. In Pop and in Rock, sometimes you don't always hear every word and it's kind of okay if it's buried just a little bit, but Country is usually not that way. People defi-nitely sing along with Country songs, so that's the biggest thing. The vocal rules. But at the same time, it's pretty boring if it's all vocals. It sounds like a Country record from the 60's where you don't have any power in there. There's an art to keeping the vocal on top without making it dominate.

*What changed and how does that affect what you do?*
Back when I used to listen to my dad's old Ray Price and Jim Reeves Country records, they weren't very far from what Pop was in the early 60's. Very mellow, big vocals, very subdued band, very little drums, strings, horns, lush. Mix-wise, there wasn't really too much difference in an Andy Williams record and one of the old Jim Reeves records.

What happened was that Country got too soft sounding. You'd cut your track and then do some sweetening with some horns and strings. At one time strings were on all the Country records and then it kind of transformed into where it's at today, with almost no strings on Country records except for big ballads. For the most part, horns are completely dead. They're almost taboo. Basically it's rhythm track driven and not really very far off from where Pop was in the mid-to-later 70's. The Ronstadt "It's So Easy to Fall in Love" and "You're No Good" where you hear guitar, bass, drums, keyboards, a slide or steel and then a vocal background, that's pretty much the format now, although fiddle is used also. Ironically enough, a lot of those guys that were making those records have moved here because at this point, this is one of the last bastions of live recording.

*Let's talk about your mixing. When you start to mix, how do you build it?*
Well, I'll usually go through and push up instruments to see if there's any trouble spots. All this is dependent upon whether it's something that I've recorded or if I'm hearing it fresh and have no idea what it is. If that's the case, then what I'll do is kind of rough mix it out real quick. I'll push it up and see where it's going before I start diving in.

If it's something that I know what's on the tape, then I'll go through and mold the sounds in a minor way to fit the modern profile that it needs to be. In other words, if it's a real flabby, dull kick drum, it doesn't matter what the vision is. This kick drum's never going to get there. So I'll pop it into a Vocal Stresser or I'll do whatever I have to do. I'll work through my mix like that and try to get it up into the acceptable range, or the exceptional range, or at least somewhere that can be worked with. It takes a couple of hours to get good sounds on everything and then another couple of hours to get real good balances, or something that plays itself as if it makes sense. Then I'll do some frequency juggling so that everybody is out of everybody else's way.

The tough part, and the last stage of the mix, is the several hours it takes for me to make it sound emotional and urgent and exciting so that it's just … not a song, it's a record. It's not making it just sound good, it's making it sound like an event. Sometimes that means juggling the instruments or the balances or adding some dynamics help. That's the last stage of when I mix, and that's the part that makes it different or special.

*So how do you go about doing that?*
I try to find what's important in the mix. I try to find out if the lead vocal is incredibly passionate, then make sure that the spotlight shines on that. Or if the acoustics are sitting there but they're not really driving the thing and they need to. If for instance, if the mix needs eighth notes, but they're going [*sound effect*] and it's not really pushing the mix, sometimes playing with compression on the acoustics or auditioning different kinds of compression to make it sound like, "Boy, this guy was into it." Maybe pushing and pulling different instruments. Somebody's got to be back and sometimes it's better when things are back and other things are further up front. It's just basically playing with it and trying to put into it that undefinable thing that makes it exciting. Sometimes it means making sure your cymbals or your room mics are where you can actually feel the guy, or sometimes adding compression can be the answer to making the thing come alive. Sometimes hearing the guy breathe like the old Steve Miller records did. They had that [*breathing sound*]. With a little of that, you might say, "Man, he's working. I believe it." It's a little subconscious thing, but sometimes that can help.

*When you're building your mix, are you starting with bass first, or starting with the kick drum?*
I start with the kick drum sound. But then I put up the drum kit and put the bass in. Then I'll push up all the statics that aren't going to have giant moves like the acoustic stuff, keyboard pads, maybe a synth or Rhodes or piano that doesn't have a whole bunch of stepping out licks. Early on, I'll try to make sure that there's room for the lead vocal. I think one of the big mistakes is to work on your track for eight hours and get it blistering hot and barking, but there's no way that this baritone vocal can cut through. So then you're forced with the choice of turning this baritone vocal into steel wool with ridiculous EQ or just turning him up so loud that he sounds inappropriate. It's cool to have a bright record as long as everything kind of comes up together, but if you've got an incredibly bright snare drum and the vocal's not so bright, then it makes the vocal sound even duller. If you are thinking all the way to the end, to when you add EQ when

you master the record, it'll brighten the vocal but it's also going to bring up the snare worse. So you have to have everything in perspective.

But eventually, I get the vocals in and get the backgrounds around them. Then put up the solos and the signature stuff. Then I get an overall rough balance of everything that sits there pretty well and then juggle the pieces. Once again, it helps if I know what the music is, then I know exactly where I'm going. If I don't, sometimes I have to listen to a rough mix or create a rough mix on the board to get a feel for what their intent is.

*Do you have a method for setting levels?*
Usually a good place to start is the kick drum at –6 or –7 or so. I'll try to get a bass level that is comparable to that. If it's not exactly comparable on the meter because one's peaking and one's sustaining, I get them to at least sound comparable. Because later in mastering, if you affect one, you're going to affect the other. So as long as the ratio is pretty correct between the two, then if you go to adjust the kick at least it's not going to whack the bass way out as long as they relate together. That's kind of a good starting place for me.

I used to let the snare dominate real hard and heavy but now I'm pulling back just a little bit instead of bludgeoning the audience. I'm letting them get into some of the other midrange things.

*Do you have a special approach to EQ?*
I don't know if I have a special approach. I just try to get stuff to sound natural, but at the same time be very vivid. I break it down into roughly three areas: mids and the top and the bottom. Then there's low mids and high mids. Generally, except for a very few instruments or a few microphones, cutting flat doesn't sound good to most people's ears. So I'll say, "Well, if this is a state of the art preamp and a great mic and it doesn't sound that great to me, why?" Well, the mid range is not quite vivid enough. Okay, we'll look at the 3k, 4k range, maybe 2500. Why don't we make it kind of come to life like a shot of cappuccino and open it up a little bit? But then I'm not hearing the air around things, so let's go up to 10k or 15k and just bump it up a little bit and see if we can kind of perk it up. Now all that sounds good but our bottom is kind of undefined. We don't have any meat down there. Well let's sweep through and see what helps the low end. Sometimes, depending on different instruments, a hundred cycles can do wonders for some instruments. Some-

times you need to dip out at 400 cycles, because that's the area that sometimes just clouds up and takes the clarity away. But, a lot of times adding a little 400 can fatten things up.

On a vocal sometimes I think, "Does this vocal need a diet plan? Does he need to lose some flab down there? Or sometimes we need some weight on this guy so let's add some 300 cycles and make him sound a little more important." So it's kind of contouring.

Also, frequency juggling is important. One of the biggest compliments people give me is that they say, "You know, Ed, on your mixes, I can hear everything." There's two reasons for that. One is, I've pushed things up at the right time or the right things up that they want to hear or need to hear. But the other thing is, you don't EQ everything in the same place. You don't EQ 3k on the vocal and the guitar and the bass and the synth and the piano, because then you have such a buildup there that you have a frequency war going on. So sometimes you can say, "Well, the piano doesn't need 3k, so let's go lower, or let's go higher." Or, "This vocal will pop through if we shine the light, not in his nose, but maybe towards his forehead." In so doing, you can make things audible and everybody can get some camera time.

*Do you have a specific approach to panning?*
Yeah, I do. The most significant approach is I pan as if I were sitting in the audience, especially with the drums. The reason is, I don't play the drums, therefore I sit in the audience and listen and that means with most drummers (unless they're left handed) put their hi-hat to the right. To me, I can get away with anything except the drums being backwards because it just strikes me funny. So I do the drums that way. However, I thrash a bit at piano so I always put the low end on the left hand side and the high end on the right hand side.

*Hard left and hard right?*
Usually, but not always. With a piano, it depends on who recorded it and how phase coherent it is. If it's not dramatic stereo, I'll try to make it more dramatic. Also, if whoever recorded it didn't pay real good attention to the phasing on the mics and the thing is way wide and it falls apart in mono, I'll be panning it in so that in mono it doesn't go away. Sometimes flipping the phase on one side can fix that because a lot of people don't check. Of course stereo is more important now

than ever before, but on a lot of the video channels, you're listening in mono. So, I check for that.

I always put the electric guitar on the left and steel on the right. I try to make stereo records and I'm not afraid to pan something extremely wide. I like my mixes to have a few things that stick out and get some attention and not just blend in with the crowd. That way, there can be all kinds of contrast, not only volume dynamics, but panning dynamics as well.

One of the things I don't like is what I call "big mono" where there's no difference in the left and the right other than a little warble. If you pan that left and right wide, and then here comes another keyboard and you pan that left and right wide and then there's the two guitars and you pan them left and right wide, by the time you get all this stuff left and right wide, there's really no stereo in the sound. It's like having a big mono record and it's just not really aurally gratifying. So to me, it's better to have some segregation and that's one of the ways I try to make everything heard in the mixes. Give everybody a place on the stage.

*How about compression? Do you use it as an effect, or just to even things out, or both?*
Both. I have a lot of different compressors, and one of the reasons I have a lot of outboard gear is because they're all different colors on the palette. A dbx 165 is basically working all the time, but you can't really hear it working and you can't really get it to suck and blow. If you want the suck and blowish thing, there's several other ways to go. An 1176 or one of several VCA compressors can really do something dramatic. It also depends if it was cut with compression. Sometimes it doesn't need any or as much, and sometimes you need it to give it life.

To me, the key to compression is that it makes the instrument sound like it's being turned up, not being turned down. If you choose the wrong compressor or you use it the wrong way, then your stuff can sound like it's always going away from you. If you use the correct compressor for the job, you can make it sound like, "Man, these guys are coming at you." It's very active and aggressive. Quite often, I'll use it on the stereo buss but I try not to be too crazy with it.

But if you remove all dynamics or if you really lean on it in an improper way during mixing, when it goes to mastering there's not much for the guy to do there. If he does, it'll only compound the problem. Then by the time it gets on the radio there's nothing left that'll pump the radio compressors, so then it just kind of lays there. It's loud, but nothing ever really jumps out of your mix. So nothing ever gets real loud.

But yeah, lead vocals almost always. Bass, certainly when I'm tracking it, and quite often when I'm mixing it. I time the release to the tempo of the song or to the end of the note release, especially if the guy's using flat-wound strings for more of a retro bass that has a lot of attack and less hang time. Sometimes if you use the wrong compressor on a snare drum, all you'll get is the initial [*sound effect*] and then it'll turn it down, but if you use the right kind of compression, slow the attack down, speed up the release maybe, you'll get a different effect. There's more length to the snare type of a sound. It'll come sucking back at you. Compression's important, but it's gotta be the right kind and I think that's the key.

*How about gates? Do you use them often?*
Well I do, but I'm not fanatical about gates. There's two reasons to use gates. One is to get rid of amp hiss or something that's not attractive to the music. Or if the hi-hat's wiping out your snare sound that leaked in on your snare mic because you added a bunch of EQ, then the gate can help you. But generally we're on digital tape and the sounds are done in different rooms so it's not as important to me. Now one exception is if I'm doing my room mics that I was talking about. To have the room mic hang up there with all that kick drum, that can wipe out your kick drum sound per se, and it also makes it lean towards Led Zeppelin, which might be the red flag. "Ooh, what are we doing here, sounds like a Rock song and it shouldn't." If it should, great. But if it shouldn't, then a lot of times what I'll do is time the gate to the tempo, so that the kick drum's out of the way. Open/close, open/close, and then play with the ramp so that it doesn't just sound trendy.

*How about setting up effects? Do you add effects as you go along or do you get the balance up and then add the effects?*
Well, I kind of come and go with this. I'm in a drier phase now than I used to be. What I'll do is try to make things sound as good as I can dry. If I hear something that just sounds too iso'ed and too unrelated to the mix, then I'll add some effects as I go, but I don't go too crazy with it until I get the whole picture. Then once it's all sitting there, you can easily tell if it's just not gluing together. My general setup for a mix is I'll have one send setup for long verb and another setup for a short, kind of a room simulation.

*Long being what, 2.5, 3 seconds?*

Yeah, 2.5, 2.3. For a ballad, sometimes 2.6. Then I'll usually have a delay send with something, whether it's eighth note or sixteenth note or dotted eighth triplets, that kind of works with the music. Then sometimes I'll have a little pitch change, like a Publison or an AMS harmonizer kind of sound. I may have a gated reverb or something that can kind of pull sounds together. Sometimes an isolated guitar sounds great dry and in your face by itself but other times it seems like, "Wow, they had an overdub party and look who showed up." Sometimes a little of that gate can kind of smear it together and make it sound like he was actually on the floor with them.

*So what are you trying to accomplish with the effects? Are you trying to make everything bigger or give it depth?*

Sometimes depth and sometimes you just want it to sound a little bit more glamorous. Other times you just want it to sound appropriate. Well, appropriate to what? If it's an arena Rock band, then all this room stuff is going to make it sound like they flunked out of the arena circuit and they're now doing small clubs. But if you got a band where that's more of an in-your-face, hard driving thing, you want to hear the room sound.

I've done records where I didn't use any effects or any verb, but quite often just a little can make a difference. You don't even have to hear it but you can sense it when it goes away. It's just not quite as friendly sounding, not quite as warm. Obviously an effect is an ear catcher or something that can just kind of slap somebody and wake them up a little bit in case they're dozing off there.

*Let's talk about monitoring. How loud do you usually listen when you're mixing?*

I mix at different levels. I try not to mix too loud because it'll wear you down and fool your perspective. I don't find it extremely valuable to listen loud on big wall monitors very often. The only reason I'll go up there is to check bottom end. That's the best way to do it, but most of the time I work off my near-fields and I try not to get too loud.

In fact, what I like to do is use the studio bigs 1 percent of the time, my near-fields 70 percent of the time, and then use a third reference that's not straight on me, but off to the side in a different place. My philosophy is that most people don't sit right in-between the speakers when they listen to music. They're in the kitchen and the music's in the living room. Even in the car, you're off to one side a little bit. So to me, that's a valid place to arrive at kind of an average.

Sometimes it's very valuable to turn things down, but there's an up and down side to both. If you listen too soft, you'll add too much bass. If you listen too loud, you'll turn the lead vocals down too much. What I like to do is make it sound good on all three unrelated systems, then it's got to relate to the rest of the world.

*Do you ever go out and listen in the car or go in the lounge and listen through the door or anything like that?*
Yeah, although I don't go to the car as much as I used to. What I'll do about an hour before printing the mix is prop open the control room door and walk down the hall or into the lounge where the music has to wind its way out the door. I find that very valuable because it's not like hitting mono on the console, it's like a true acoustic mono. It's real valuable to see if you hear all the parts and it's real easy to be objective when you're not staring at the speakers and looking at the meters.

*How many versions of a mix do you do?*
Generally, I like to put down the mix and then I'll put down a safety of the mix in case there was a dropout or something went goofy that no one caught. Once I get the mix, then I'll put the lead vocal up half a dB or eight-tenths of a dB and this becomes the vocal-up mix. Then I'll do a mix with all vocals up. Sometimes I'll recall the mix and just do backgrounds up and leave the lead vocals alone. Then I'll do one with no lead vocal and just the backgrounds. Then I'll do one with track only, just instruments. That's usually all the versions I'll need to do. There's some people that get so crazy about it because they don't want to make a decision. At some point, you burn up $400 worth of tape or whatever you're using, just to print all these mixes. If you're going to MO, it's kind of expensive; it's $80 for 80 minutes. Generally, those cover about all of them.

*Do you always go to MO (magneto optical) or do you sometimes go to tape?*
If it's a project I'm producing or if it's a project that I'm being asked what I want to do, I'll print to MO and two DATS simultaneously and keep one mix at the studio and give one to the producer. If I'm the producer, I'll throw it in my bag so that there's a backup, assuming somebody's going to lose their DAT or something's going to happen. Some people say, "Look, we

just can't afford the MO, it's a little expensive and HDCD is expensive. Let's just go to DAT." Okay, then we'll print to just the two DATs.

The thing that impresses me the most about MO and the HDCD is that I've always wanted to play back the mix and have it sound like the mix playing off the console. [The] 1/2" tape was good, only there was tape compression that may or may not be what you want. The bottom bump was different and the hiss was different. Even if you print hot, you drive the hiss down but the peaks are different. Even some of the dig formats weren't quite right. The width would close and the verb would go away. They'd get harsh and they'd get cold with no bottom. But with this HDCD and the MO, man, it is identical. You switch back and forth and you struggle to hear the difference.

# CHAPTER 27

# Allen Sides

**A**lthough well known as the owner of the premier Oceanway studio complexes in both Los Angeles and Nashville, Allen is also one of the most respected engineers in the business, with recent credits that include the film scores to *Dead Man Walking, Phenomenon,* the Goo Goo Doll's *Iris* and Alanis Morrisette's "Uninvited" for the *City of Angels* soundtrack, as well as records with the Brian Setzer Big Band and *The Songs of West Side Story,* featuring Phil Collins, Natalie Cole, All For One, Trisha Yearwood, Wynonna Judd, Tevin Campbell, Kenny Loggins, Michael McDonald, Little Richard and Aretha Franklin.

Even though he remains on the cutting edge of the latest that recording technology has to offer, Allen continually finds modern uses for many long forgotten audio relics, proving that sound technique, good ears and an interesting piece of gear never go out of fashion.

*Do you know what you're going for before you get there? Can you hear the finished product in your head?*
It depends. I would say that if it's a project that I've been working on, I've already put it up dozens of times, I have a pretty good idea of what I'm doing. If it's something I'm mixing for someone else, then I listen to their roughs and get a concept of what they have in mind. I really want to understand what they want so I can make that a part of the picture that I draw. How many times have you had a mix that you thought was killer and they come in and change a few things and it's become perfectly acceptable, but no longer great?

*Do you have a special approach to mixing or a philosophy about what you're trying to accomplish?*
First, I like it to be fun to listen to. I'll do whatever it takes to make it satisfying. I tend to like a little more lows or extreme highs and a lot of definition, and I like it to sound as punchy as I can make it. So much involves the arrangement. When the arrangement is great, then the music comes together in a very

nice way. If it fights, then it becomes very difficult to fit together. Getting the arrangement right is an art in itself.

*How do you go about building a mix? Where do you normally start from?*
I would say that it really varies. Sometimes I'll throw up everything and then after I hear how the vocal sits, then I'll look at a section and refine it. But before I do, it's really nice to hear how it relates to the vocals because you can spend time making the whole thing sound great but it might not relate to the vocal in any way. So I'd say that I listen to the whole thing, then go back and work on each section separately, then put it all together.

*Do you have a method for setting levels?*
Yeah. When I set up my track, I'll set the monitor level to where I'm comfortable and I will make it sound as impressive as I can at maybe −2 on the VU meters because I know I'm going to come up from there. I want to make it as impressive as I possibly can at a fairly modest setting.

*This is the whole mix now.*
Yeah. I get it to where it's really kicking. Then I do my vocal track and get it all happening. Even when I do that, I probably will end up trimming the individual faders here and there. The problem, of course, is that when you trim the individual faders, the way that they drive the individual effects changes slightly. All the plates and effects sound different when they're driven differently. That's why I try to get everything happening in that lower level so I have to do as little trimming as possible. And I like to keep my buss masters all the way up. This, of course, depends on the console.

*So you're putting the whole mix up first and then you're adding the vocals later.*
Yeah, but as I say, I will probably put the whole mix up, put the vocals in, and listen to how it all fits together before anything. Based on that, I think it's a decision of how I'm going to make the rhythm section sound.

And another thing I'd say is that I'm definitely a fan of your first impression being your best impression. I like to move very quickly so no matter how complex it is, within two to three hours it's kind of where it should be. Some of the mixes are so complex these days that you have three and four 24-tracks locked together, or two 48s and a 24. It's insane. So a lot of times the music is so complex that you can't actually hear the mix until you put all the mutes in with all the parts playing in the right place. If you just put all the faders up then you'd have one big mess. So there's a tremendous amount of busy work just to get it prepped so that you can play it back.

*Do you have an approach to the rhythm section in particular?*
Believe it or not, I typically bring in the overheads first because my overheads are overall drum mics. So I bring in the overheads to a good level, then I fit the snare into that. Then I get the kick happening. Then I take a look at the toms and see if they're really adding a lot of good stuff to the drum sound. I'll just keep them on and set them where I want and then push them for fills. If they tend to muddy things up then I'll do a whole set of mutes so they're only on for the tom fills. Obviously you can set certain ambience and effects on the toms that you don't want on the rest of the kit, and you can make them as big as you want that way. I hate gates. I'd much rather control every fill myself. But usually overheads first, then snare, then kick and then the toms; see how it fits, then tuck in the hi-hat.

*Do you have an approach to EQ?*
What I would say is that I tend to like things to sound sort of natural but I don't care what it takes to make it sound like that. Some people get a very preconceived set of notions that you can't do this or you can't do that. Like Bruce Swedien said to me, he doesn't care if you have to turn the knob around backwards; if it sounds good, it is good. Assuming that you have a reference point that you can trust, of course.

Here's a good example of that. Do you know what UA500 EQs are? It's a passive EQ with a great sounding 15k. I remember one time I was doing a record and I had a really great sounding Steinway B but it was very, very soft. I think I boosted 10dB at 15k on the left and right channels, which sounds exorbitant but it sounded completely natural. A tastefully EQed piano played softer can sound better than a bright piano.

When I'm recording digitally, particularly when I'm printing things that are fairly quiet and fairly soft, I want to make sure that my harmonic structure above 12kHz has sufficient energy to be in a higher bit range. Because if you have to boost when it comes back, there's nothing there but noise and grit. So I think that I tend to do most of my EQing when I record. That's often the opposite of what many people do, but I'm very careful and everything is to taste so that it sounds good. I really hate mixes where they've taken various center points of the band and boosted the crap out of them. The harmonics are gone and all you end up with is just harsh unpleasantness.

*Do you add effects as you go along, or do you put the mix up and balance it and then add the effects?*
No, the effects are usually added as I go along because a lot of times I'll work on multiple image effects on kicks and snares and stuff and tie that in to overheads so you can hear all the sounds as a single entity. Obviously that can change again when the vocals come in. Invariably what works by itself is not going to be exactly the same when you put the vocals in. You may have to increase or decrease those effects to get your overall picture to happen.

The other important thing is that when I'm using effects, I hate it to sound generic. I'd much prefer it almost to sound like we're going for a room sound. You have a great natural kick and snare, plenty of attack and punch and the ambience surrounds it in such a way that it doesn't sound like an absolute tin can cheese-ball effect but becomes more of a natural sound. Obviously, it's relative to the music you're doing. It's all different.

*And so that's what you're trying to do then more than anything, trying to get something that's more of an ambient type of thing?*
Yeah, there's also a question of dryness versus live-ness versus deadness in regards to monitor volume. Obviously when you turn it down, your ambience determines how loud it sounds to you to some degree. And if you're monitoring at a loud level and it's very dry, it can be very impressive sounding. When you turn down, it might not be quite so full sounding so obviously there's a balance there. I would say that I'm getting dryer these days.

*Seems to be the trend, actually.*
I'm definitely getting back there, because I started in the late 60's R&B days which were very dry.

*Do you have a method for setting up your delays and reverbs?*
Yeah, I'm a big fan of tube stereo plates. I've got 25 plates here. And then I use this old Ampex mastering preview DDL for my pre-delay to my chambers. I have about nine or ten of these and they're the greatest DDLs on the planet. Then I usually have API EQs across the sends.

I have some really great live chambers too. I'm a big fan of the RMX16, not for drums, but for vocals and guitars and stuff. I love Non-Lin for guitars and things. Let's say that you had a couple of discrete guitars that were playing different lines and you try putting them in the middle and they get on top of each other. If you put them left and right, they're too discrete. The RMX Non-Lin set at 4 seconds with a 10 millisecond pre-delay and an API EQ on the send with about +4 at 12k shelf and –2 at 100Hz going into it does a wonderful job of creating a left/right effect, but it still spreads nicely. It works great for that.

*Do you have an approach to panning?*
Yeah, I tend to do a lot of hard panning [*laughs*]. I don't pan in much. I am really big on having things wide. Obviously the reason for panning in is because there's tremendous value in returning to mono, particularly in reverb returns. I still do a lot of comparing between mono and stereo. No matter what anybody says, if you're in a bar, you're going to hear one speaker. There still has to be a relevance between the stereo and mono thing.

I tend to like things very wide. I just think you can't make it wide enough because I find that one of the biggest problems I have with the current digital formats is losing width. It never gets bigger, it only gets smaller and teenier and tinier. So I work so hard to use every possible bit of information I can and it makes such a difference.

*How about compression, do you have an approach to that?*
Sometimes I use our Focusrite [console] setup, which has three different stereo busses that can combine and take a mult of the initial totally clean program and nail it to the wall to bring up all the little ambient stuff and just tuck that back into the main clean buss so that you can add this sustain that everybody wants without killing the attack. If I take one of my SSL limiters and do that thing that it does, of course it always suffers from a certain lack of impact. So a lot of times we want to get that sustain, particularly on a rocking track, but still want a hell of a punch. That's a way to do that.

I virtually never limit basses during the initial recording process. With players like Nathan East and Jimmy Johnson, who are so consistent and whose instruments sound so good and so well balanced in the track, all limiting would do is mess it up. But sometimes when you have a player in a band that's somewhat inconsistent and you need to bring it into balance, tasteful compression can be helpful, but you still want to leave dynamics. That's often times one of the worst problems I encounter. There's no dynamics left, it's all gone.

*Squeezed to death.*
There's no impact, so I end up having to do as much as I can to put it back in. There's this box that isn't made any more called the Marshall Time Modulator. You can set the Marshall Time Modulator so it's just on the verge of feedback, take a bass that's been limited to death, and it's possible to get the Marshall to expand the peaks. You set it so it's just on the verge of feeding back and it'll actually push those peaks out. Amazing box. It's the only device I've ever seen that'll do that.

Have you ever used the Lang? Say you pull up the overheads and there's no highs above 8kHz and it's harsh because basically it's been recorded on the 24-track at +18. So if you take a nice set of EQs, maybe a GML or some APIs and boost the top end, it just gets harsher and doesn't really add highs. If you take a Lang, set it at 15k, then set it as narrow as the bandwidth will go and boost it, all of a sudden there is this silky 15k top end. If you set it narrow enough, it just ring modulates and adds a beautiful silky 15k harmonic that did not exist on the program you boosted.

Or, say that you've got a vocal and you want to get that little air thing, but no matter what you try to do, it just gets harsh. The Lang is amazing. It's one of those boxes where you get something for nothing.

*Do you have a monitor setup that you usually use, and what level do you listen back at?*
I must admit that I really do enjoy big speakers. I like to turn it up and have fun. I have no problem mixing on anything else, but I like having nice accurate big speakers that are fun to listen to, that aren't harsh and that don't hurt my ears.

Generally speaking, when I put up the mix, I'll put it up at a fairly good level, maybe 105, and set all my track levels and get it punchy and fun sounding. Then I will probably reference on NS10s at a very modest level just to check my balance and go back and forth. The small speakers that I'm fond of now, the Genelecs 1032s, I can mix on totally without a problem. But I love my big speakers and I have so much fun [*laughs*].

And if I listen loud, it's only for very short periods of time. It's rare that I would ever play a track from beginning to end loud. I might listen to 20 seconds or 30 seconds of it here and there, but when I'm actually down to really detailing the balance, I'll monitor at a very modest level.

*Modest meaning how loud?*
I would say that at a level that we could have a conversation and you could hear every word I said.

*Do you have any listening tricks like going to the back of the room, outside and listen through the door...*
Oh yeah, I think we all have that. You walk out to get a cup of coffee and you're listening to it outside in the hall and the realization strikes you, "How could I miss that?" because it's a different perspective. What I love is my car. With the automation we have today, I put it on in the car on the way home and any additional changes I hear, I just call them in and have the seconds print the updates.

*How often do you have to go back and redo your mix or touch it up?*
I would say that about ten percent of the time I have to go back and do that, so I have my guys document every single detail as best as I can. To be honest with you, sometimes it comes back and sometimes it doesn't because even the different digital devices are different if it's not the same unit. And if it's a song that really relies on a particular slap or effect or harmonizer or something that is really a major part of the ambience of the song, a half dB level in an aux send is going to change the whole musical balance of a song. It just kills us to have a mix that we really are happy with and they say, "I love the mix, just change one thing," because they just don't understand. And then of course, the joke of all time is to mix on an SSL or a VR in any given room, and then go to another room and put it up. Well, forget it. It's one thing putting it up on the same console, but the accuracy of the pots are so different from console to console.

*How many versions of a mix do you usually do?*
Plenty. Invariably I will do the vocal mix to where I'm totally happy with it and then I'll probably do a quarter and half dB up and a quarter and half dB down. I'll print as many mixes as needed, depending on how difficult the artist is to please. Then if I need to, I'll chop in just the part I want. If there's a word or two, I'll just chop those words in. I really cover myself on mixes these days. I just do not want to have to do a mix again.

# CHAPTER 28

# Don Smith

**J**ust one look at producer/engineer/mixer Don Smith's client list gives you an indication of his stature in the industry. With credits that read like a *Who's Who* of Rock & Roll, Don has lent his unique expertise to projects by The Rolling Stones, Tom Petty, U2, Stevie Nicks, Bob Dylan, Talking Heads, The Eurythmics, The Traveling Wilburys, Roy Orbison, Iggy Pop, Keith Richards, Cracker, John Hiatt, The Pointer Sisters, Bonnie Raitt and lots more.

*Can you hear the mix before you start?*
I can usually hear roughly what it should be. I start out with the basics of a good rough mix and then I try to tweak it from there. Sometimes, I may hear something while I'm doing it, like a tape delay on the drums, that might change the character of the mix and make it turn into a different direction.

*How do you start your mix?*
Most of the time just drums and bass, then everything else. Then there were some records that I started with lead vocal then guitar and the drums would be last. With somebody like Tom Petty, his vocal is so important in the mix that you have to start with the vocal. So the vocals get roughed in, and you throw guitars around it. Then I might start back in the other direction, making sure that the drums and the foundation is solid. But I like to start with the vocal and guitar because it tells me what the song is about and what it's saying, then let everyone else support the song.

*Do you have a method for setting levels?*
Yeah, I'll start out with the kick and bass in that area (–7).
By the time you put everything else in it's +3 anyway. At least if you start that low you have room to go.

*Do you have an approach to using EQ?*
Yeah, I use EQ different from some people. I don't just use it to brighten or fatten something up, I use it to make an instrument feel better. Like on a guitar, making sure that all the strings on a

guitar can be heard. Instead of just brightening up the high strings and adding mud to the low strings, I may look for a certain chord to hear more of the A string. If the D string is missing in a chord, I like to EQ and boost it way up to +8 or +10 and then just dial through the different frequencies until I hear what they're doing to the guitar. So I'm trying to make things more balanced in the way they lay with other instruments.

*Do you have a special approach to a lead instrument or vocals?*
For vocals, just make sure that the song gets across. The singer is telling a story. He's gotta come through but not be so loud that it sounds like a Pepsi commercial. Sometimes you might want the vocal to sit back in the track more because it might make the listener listen closer. Sometimes you don't want to understand every word. It depends on the song. It's always different.

*Do you build a mix up with effects as you go along?*
I always build it up dry. I look at it like building a house. You've got to build a solid foundation first before you can put the decorations on. The same way with tracking. I very rarely use effects when I track. Just every now and again if an effect is an integral part of the track to begin with, then I'll record that.

What I've found is that if you really get it good butt naked, then when you dress it up, all it can do is get better. If you put on effects too early then you might disguise something that's not right. I don't really have too many rules about it, I'll just do what feels good at that moment. Sometimes you get it butt naked and you don't need to put any effects on. It's pretty cool, so just leave it alone.

*Do you have a method for adding effects?*
I usually start with the delays in time, whether it's eighth note or quarter note or dotted value or whatever. Sometimes on the drums I'll use delays very subtly. If you can hear them, then they're too loud but if you turn them off, you definitely know they're gone. It adds a natural slap like in a room, so to speak, that maybe you won't hear but you feel. And, if the drums are dragging, you can speed the delays up just a nat so the drums feel like they're getting a lift. If they're rushing, you can do it the other way by slowing the delays so it feels like they're pulling the track back a bit.

A lot of times in my mixes you won't hear those kinds of things because they're hidden. On the Stones *Voodoo Lounge* album there's a song called "Out of Tears." There's these big piano

*The Mixing Engineer's Handbook*

chords that I wanted to sound not so macho and grand, so I put some Phil Spector kind of 15ips tape slap on it. It sounded kinda cool, so I tried some on the drums and it sounded pretty cool there too. By the end of it, I had it on everything and it changed the whole song around from a big grandiose ballad to something more intimate. It was played on a Bosendofer but really wanted more of an upright, like a John Lennon "Imagine" type of sound.

*Do you use tape slap a lot?*
I use tape slap all the time. I use it more than I use digital delays. It's a lot warmer and much more natural and the top end doesn't get so bright and harsh so it blends in better. I varispeed it to the tempo or whatever feels right. I usually use a 4-track with varispeed and an old mono Ampex 440 machine for vocals. The mono has a whole different sound from anything else. Sort of like the Elvis or Jerry Lee Lewis slap where it can be really loud but never gets in the way because it's always duller yet fatter.

On the 4-track, I'll use two channels for stereo, like for drums, and send each slap to the opposite side. Then the other tracks I might use for guitars or pre-delay to a chamber or something. Sometimes I'll put Dolby on the 4-track to cut down the hiss or at least turn the gain way, way up cause you're not using much of it.

*Do you have an approach to panning?*
Yeah, it's kinda weird, though. I check my panning in mono with one speaker, believe it or not. When you pan around in mono, all of a sudden you'll find that it's coming through now and you've found the space for it. If I want to find a place for the hi-hat for instance, sometimes I'll go to mono and pan it around and you'll find that it's really present all of a sudden, and that's the spot. When you start to pan around on all your drum mics in mono, you'll hear all the phase come together. When you go to stereo it makes things a lot better.

*Do you have a set of monitors that you use all the time?*
I have a set of Yamaha NS10s that I've had since '83 as well as a set of RORs from '80 or '81, which they stopped making. I tried all the different versions but they never sounded the same afterwards. I bring a Yamaha 2101 amp with me sometimes.

*What level do you listen at?*
I like to listen loud on the big speakers to get started, and occasionally thereafter, and most of the time at about 90dB. When the mix starts to come together it comes way down, sometimes barely audible. I turn it down way low and walk around the room to hear everything.

I mix a lot at my house now where I sit outside a lot on my patio. If I mix in a studio with a lounge, I'll go in there with the control room door shut and listen like that. I definitely get away from the middle of the speakers as much as possible.

*How many mixes do you usually do?*
I try to just do one mix that everybody likes and then I'll leave and tell the assistant to do a vocal up and vocal down and all the other versions that they might want which usually just sit on a shelf. I'll always have a vocal up and down versions down because I don't feel like remixing a song once it's done.

*What format do you usually mix to?*
1/2" 30ips to BASF 911. I'm not into digital.

*How much gear do you bring with you?*
Quite a bit. Mostly old stuff like Fairchilds, Pultecs, API EQs, Neve compressors, 1176s, an EMT 250. I've got just a handful of anything new, like a TC5000, two SPXs and two SD3000s.

*How much compression do you use?*
I use a lot of it. Generally, the stereo buss itself will go through a Fairchild 670 (serial #7). Sometimes I'll use a Neve 33609 depending on the song. I don't use much; only a dB or two. There's no rule about it. I'll start with it just on, with no threshold, just to hear it.

I may go 20:1 on a 1176 with 20dB of compression on a guitar part as an effect. In general, if it's well recorded, I'll do it just lightly for peaks here and there. I'll experiment with three or four compressors on a vocal. I've got a mono Fairchild to Neve's to maybe even a dbx 160 with 10dB of compression to make the vocal just punch through the track.

Again, I don't have any rules. As soon as I think I've got it figured out, on the next song or the next artist, it won't work as well, or at all.

# Guy Snider

A former guitar player who played with the likes of Ike and Tina Turner and Chuck Berry, Guy Snider's projects as an engineer run the musical gamut from rockers Nine Inch Nails and Faith No More to the smooth R&B of Brandy and Shante Moore to hard-core rappers Tupac, Snoop Doggy Dog and Nate Dog. Fresh from a Rolling Stones project, Guy talked about being a Rock & Roll kind of guy in a Rap world. It should be noted that this interview was done before the untimely passing of Tupac Shakur, hence the reference to him in the present.

*What's different about doing Rap as compared to every other genre of music?*
It's very unlike a normal production the way Rap, especially the Death Row stuff, is done. In most Pop music the chain of command is the artist, producer, then engineer. Here is the song, and the producer and the artist agree on the arrangement, they cut it and the engineer mixes it until all parties are happy with it.

In Rap it's not so clearly defined. For instance, there'll be a producer who is generally the programmer and will generally use a sequencer/sampling machine like the Akai MPC3000 to take a sample from an existing song. For instance, there was a song by Sting called "Mad About You" off the *Soul Cages* record, where we sampled the beginning of it and looped it into a four bar thing. That'll be like the melody. They then will add kick drums and 808 sounds internally in the machine and call that a "beat." Then Tupac will hear it and write lyrics to it, and go out and rap to that. Various guitar players are then called in at that time, so it's very spontaneous. So you might walk out of the studio thinking that you engineered and mixed something, but then the next night without you knowing it, somebody else will come in and add another thing to it. And then somebody else will put a mix to it.

By the end, they'll have eight or nine different versions of the song, so by the time the album gets released, who produced, engineered and mixed which cut is uncertain because so many different people were involved at so many different times over so many different versions.

*How is mixing Rap different from other types of music?*
In Rap, the main thing is the beat and the lyric. Sonically it has to have that big bottom and make you want to groove, but lyrically, the enunciation is very important so you have to be very careful about adding effects and delay. For instance, if you're doing an Annie Lennox thing, you have all this ambient sound going on and you can add those types of processing to her voice. It's OK if the sound of her voice blends in with the music, even if you lose articulation of the lyrics, because you're going for the sonic effect. When you're mixing Rap, you have to be aware where you go sonically with your instrumentation because you can't just have this wet mix against the dry vocal. Also, recalling exactly what you had a week before is imperative so we have to keep the outboard processing to a minimum, since most pieces sound different from unit to unit and it's really hard to get the exact sound back.

*What would you consider your approach to mixing? What are you trying to accomplish?*
To have the sounds touch an emotional response. And it's not necessarily in the lyric or a musical note, but the way it's placed to touch an emotional response. When I study other engineers I respect, I generally do not go, "Wow, I like their kick or snare sound." I like their interpretation. Something like having background vocals come in on one side and then having an ad lib on the other side. To me, that's the stuff that makes records exciting. Anybody who's gone through recording school can get sounds, but the interpretation is what separates the men from the boys.

*How do you build your mix? Where do you start from?*
I throw up all the faders and listen to what the song is about, and then I start with drum sounds. Then I throw the vocal in. The vocal's always sort of the barometer. Plus I use it also for frequency placement. A lot of times the snare is fighting for the same space as the vocal, so I always make sure that the snare sounds really good when it's loud and popping in your face but also sounds good pushed back into the mix. Also, a lot of times in Rap there'll be breakdown sections, so the song should really carry itself with only the drums, bass and vocals.

*Are there certain EQ points that you seem to come back to all the time?*
Yeah, on kicks I always dip 800Hz out. It depends, on Rock
either the kick drum carries the low end of the song or the bass
guitar carries the low end of the song. On Rap, there might be
two or three low end things going on. One thing that annoys the
heck out of me, especially in Rap, is when they turn up the
bottom so it becomes just a rumble. For me, the low end has to
be defined.

The trick I do is to take the kick and the snare and buss them
to either a stereo compressor or two mono compressors linked
together and bring that back into two channels of the console.
Then on these two channels, I'll suck out all the mids and add
tons of low end with lots of compression. Now I have two
faders that I can add to get the sub thing going, but it's very
controlled. So, if they want more low end I can unmute those
and all of a sudden they're satisfied. That's one thing with Rap,
I'm really aware that they are into the low end for low end's
sake. I like low end, but it should be in proportion to the rest of
the audio. Definition is very hard to do on the Rap thing. That's
one thing I fought with Tupac on and I fought with Snoop on.
Anybody can just add EQ to it, but getting the low end with
definition is the art.

*What level do you monitor at?*
I monitor extremely soft, to the point that assistants have to
leave the room. I have a tendency to pick up nuances at conversation volume. Like if you were in the room talking for any
length of time, I'd either ask you to leave or I'd turn up the
volume all the way in the mains until it shut you up because I
can't have conversation in the room while I mix.

*Do you bring your own monitor system?*
That's one thing that's really improved my mixing in the last
three years. I used to bring my own speakers and have this cool
stereo system at home that used Carver amplifiers going to a
pair of Tannoy Golds. I would come home and pop my DAT in
and listen to my mix. My wife at the time bought like an $80
boombox so I wouldn't wake her and my kid when I got home
late at night. So I started making cassettes instead of DATs and
going home and playing them at low volume on the boom box.
I flipped it in and out of the radio and I was really sort of
freaked about what I was hearing. I'd turn on KFCA [an LA
college station] and I'd listen to a Shawn Colvin mix and I'd
put mine on and go, "Whoa, what's going on?"

So then I started gearing my mixes towards the boombox, and I quit taking my Tannoys to the studio and quit using the KRKs and started using the old Yamaha NS10Ns for everything except getting my drum sounds. All of a sudden, the boombox tape starting sounded better and better and better and better. And you know what, the last two years I very rarely listen to my mixes two or three months later when you're removed from them and go, "Hmm, what was I thinking?" I'm now going, "Hey, I did a good job on that." I still love to listen back on a pair of Tannoys, but for the most part I stay on those little Yamahas.

*Do you find that your background doing other types of music helps you when doing Rap?*
I really never want to be labeled as a Rap engineer, or a Rock or Country-Western engineer for that matter. I think it's like getting stuck on one type of food, like saying "All I eat is Italian." To me it all comes down to that old cliché, "There's only two types of music: good and bad." Even though I'm pretty technical, I think my background as a musician helps me more than anything. And I think that's one thing you've got to remember. The knobs and all the LEDs in the studio look really cool, but none of it makes one note of music. You've got to know the technical stuff but then forget it and just listen to the music.

# CHAPTER 30

# Ed Stasium

P roducer/engineer Ed Stasium is widely known for working on some of the best guitar albums in recent memory (including my own personal favorites by the Smithereens, Living Color and Mick Jagger), so I was really surprised to discover the total breadth of his work. From Marshall Crenshaw to Talking Heads to Soul Asylum to Motorhead to Julian Cope to the Ramones to even Ben Vereen, Ed has put his indelible stamp on their records as only he can.

*Do you have a specific approach when you sit down to mix?*
Unlike some other people who are specifically mixers, I've been fortunate in the fact that everything I produce I've been able to follow through on it with the mix. I'm a "hands on" kind of producer/engineer guy.

*Where do you start to build your mix from?*
I put the vocals up first and then bring in the bass and drums. I bring up the whole kit at the same time and tweak it, but I'm not one to work on the kick drum sound for two hours. Also, I've recorded everything, so I know what's there and don't have to mess around much with anything.

*How long does it take usually?*
I would say maybe between six and ten hours. I don't use a lot of effects. I use an EMT plate and a slap tape, but everything that you hear on the mix is basically like what's on the multitrack. I consider my technique very old school. I don't use a lot of digital reverb. If I use any kind of outboard gear, it's a Pultec or a LA-2A or LA-4A or a Fairchild or even a Cooper Time Cube, that type of thing. I do use Drawmer gates on the reverb returns, just to keep them quiet.

*Do you have an approach to using EQ? Do you find any frequencies that you always seem to be coming back to, like on a kick drum?*
No. I approach it pretty haphazardly. I don't have any rules really. I just sort of move the knobs. I'd actually rather move the mic around and find whatever sounds good instead of resorting to extreme EQ.

*Do you have an approach to panning?*
My mixes are kind of mono, but not really. I pan tom-toms but not to extremes, usually between 10 and 2:00. Usually I have the drums in the middle, vocals in the middle, solos in the middle. I do pan out the guitars, though. If there's one guitar player, I'll do a lot of double-tracking and have those split out on the sides. But if there's two guitar players, I'll just have one guy over on the left, one guy on the right. And if there is any double-tracking on any of those, I'll split them a little bit but I never go really wide with that.

*Do you use a lot of compressors?*
I like compression. I think of compression as my friend. What I do a lot is take a snare drum and go through an LA-2, just totally compress it, and then crank up the output so it's totally distorted and edge it in a little bit behind the actual drum. You don't notice the distortion on the track, but it adds a lot of tone in the snare, especially when it goes [*makes an exploding sound*]. Actually, something I've done for the last 20 years is to always split the kick drum and snare drum on a mult and take the second output into a Pultec into a dbx 160VU and into a Drawmer 201 gate. Then I pretty much overemphasize the EQ and compression on that track and use it in combination with the original track.

*How about effects? You say you don't use many, but you obviously use some. Do you get your mix up first and then add everything, or add effects as you go?*
As I go. I usually have a couple of EMT140s to use. I always have a slap tape going on that I put in time with the tempo of the song.

*Is the slap for an individual track or is it specifically for pre-delay for the chambers?*
It's usually on vocals. I always have a little bit of a slap on the vocals and I might send some of that slap return to a chamber as well.

*Do you have monitors that you take with you?*
I have these little Aiwa speakers that I bought in 1983 in Atlanta. I always get my balances on those. I really like to listen at very low levels. Sometimes I try to have a pair of old JBL 4311s, 4312s, or 4310s because I still actually have the home version L100s in my house. And then there's always the Yamahas hanging around.

*Do you have any listening tricks, like going out in your car?*
I have this little stereo system I carry around with me with these Advent wedges called Powered Partners that are AV speakers. I have a little road case for them, and I bring them with me when I stay in hotels out of town. I do all my listening, even when I go home at night, on these.

*Have you noticed any changes over the years in the way things are done, or the way you do things?*
Yeah, especially with the onset of computer. I do a lot more riding of things, especially the vocal, and doing little different effect changes. It makes life easier in a way, but then it makes life more complicated because you can do so much more. It depends again on what you're doing. The Living Color records were very complicated. We had a lot of different effects on the verse, different effects on the vocal, that kind of great stuff. When I mixed "Midnight Train to Georgia" back when I first started, we did that on a little 16-input, 16-track in somebody's basement in New Jersey. The drums were all on one track and you just made sure you got the vocals right.

I remember the tracks were really packed on that song, so I just brought things in gradually. We started off with the piano, added the guitar and added the Hammond. But now, I'm riding every snare drum hit to make sure it cuts through, every little guitar nuance, little cymbal things, and the kick in certain places. I'll be riding everything.

*"Midnight Train" sounds so clean…*
That was a great console, a Langevin. I don't know whatever happened to it. I don't know where it came from, but it was in Tony Camillo's basement studio in New Jersey that we recorded that stuff on. The vocals were done in Detroit. I'm sure the drums were only on one track or two tracks at most. The Pips were double-tracked. You know, Gladys is right up in the front. We didn't use many effects on that because we didn't have any effects. It was a little basement studio and all we had was a live chamber that was the size of a closet that was concrete with a

speaker in there and a couple of microphones. That was the reverb on that record.

Same thing at Power Station. I was mixing the third Ramones record (which was actually the first thing mixed at Power Station) while we were still building the place. We had that 910 Harmonizer, a couple of Kepexes, and no reverb at all. What we used for reverb on that whole record was the stairwell.

*How many mixes do you usually do?*
I'll do a vocal up. Sometimes I do guitars up. It depends on what players are in the room. If the drummer's in the room, he'll say "Hey, can I have more snare drum?" I'll say, "Oh yeah, we'll do an extra mix with more drums in it." And if the guitar player's in the room, he'll say, "I need to hear the guitars a little more." I'll say, "Okay, we'll put the guitars up," but I always use the real mix anyway. Just kidding everybody!! [*laughs*] It doesn't matter. You get so critical when you're mixing and when it comes down to it, it's the darn song anyway. As long as the vocal's up there, it will sound pretty good. You won't even notice the little things a month later.

# CHAPTER 31

# David Sussman

A lthough mixing any type of music requires similar skills in terms of balance, equalization, panning and effects, each style has its own particular quirks that only experience in that genre can teach. One of the more intense and creative categories is Dance music, one that engineer/mixer/ producer David Sussman knows well. While engineering for producers David Morales and Grammy Award-winning Remixer of the Year Frankie Knuckles, David has developed quite a resume, engineering remix work for such artists as Mariah Carey, Whitney Houston, Janet Jackson, Madonna, Tina Turner, Gloria Estefan, Seal, Michael Jackson and U2 as well as recent additional production and mix credits for Salt-N-Pepa and MLF.

*What's the difference between mixing for Dance and mixing for Rock or R&B?*
One thing is that the arrangements are very rarely finalized. In other words, there's going to be 46 tracks of music and vocals to mix and sometimes within those tracks there could be two or three different songs. You'll have two different bass lines with their own accompanying music and neither one may be related to the other except for the vocals. So when you're engineering these kind of records, you have to know how to decipher it if you've not done the overdubs yourself or if it's not marked on the track sheets by a competent assistant or recording engineer.

Another interesting thing about doing Dance remixes is the prep work that's involved. You basically get a multitrack with the original production on it and you have to extrapolate a click and a sync. If you get a production that's at 107 beats per minute and the producer wants to make it 127, now you have to time stretch the vocals, keeping the same pitch but at the tempo where they want it. I usually try to get all the tracks that are required for the new production on one 24-track analog tape machine, but I've done it on 48 tracks as well. First I create a click track, if there isn't already one, and print it to one of the open tracks. If there's not an open track, then I'll discuss with the producer which track to burn and then I'll print the click over that track. Then we'll varispeed the multitrack until we have the right tempo and do a quick vocal mock-up through the Lexicon 2400, which is really a glorified harmonizer.

*Is that what you're using for time compression?*
Yeah. Then we'll give a quick listen and see how the vocals play through it. If they sound too much like chipmunks, then we'll bring the tempo down. If they sound fine, then that's the new tempo. We notate the varispeed of the tape machine and then find another track to burn fresh time code on. Now we have a click with time code, which can be used to educate a synch box. We then patch the vocals through the Lexicon, two tracks at a time, to a slave not running in varispeed, which is locked using the time code. That way, when you're finished throwing all the vocals over through the Lexicon time compressor, that slave tape is now our new master playing at 30ips at the right tempo.

*How long does it take you to do it?*
I would say the quickest we've done it is probably like two to three hours, but it could take us as long as five to six. Keep in mind that this is the way we've been doing it for about ten years. Lately we've been using some computer-based systems but have found this way [sounds] best when pushing the limits.

Another major thing about Dance music is that the kick drum is ultra important. You have to be able to feel it. On a lot of modern contemporary Rock records, you don't necessarily feel the kick drum, you just have to hear it. It's not going to hit you the same way as the Dance records. In Dance, it's really important to have that kick drum and bass really dry and in your face.

*What do you do to get that?*
A lot of it's in the balance of the elements in the track. A kick drum almost always goes through like a Pultec into maybe another kind of EQ, depending on what it is you're looking for. I really like the SSL on-board compression for Dance music, but I'm so accustomed to the SSL that I can basically mix a record with my eyes closed. I've tried mixing a Dance record on a Neve but found it a little difficult. The SSL has that thing for the kick and the snare compression that I like. The dynamics are sweet. Not that it works on all genres, but for Dance it's really hot.

*Where do you build your mix from? Always from the kick?*
Yeah, I pretty much start with the drums. I like to get the drums and the bass into the record within the first hour, then I'll start putting in the keyboards. Sometimes I'll do drums and vocals

and then put the keyboards in after that. If it's a really important vocal record, like for Mariah Carey or someone like that, I'll do the vocals pretty early only because I like to make sure that the reverbs for the vocals are right. I don't want to tailor all my reverbs for drums and keyboards then suddenly find that I'm out of really good reverbs for vocals. A lot of times you don't really need too much reverb on the keyboards anyway. Sometimes I'll use delays to make it sound a little wet as opposed to actually using reverb.

*Delays are timed to the track I suppose.*
Oh yeah. In Dance music especially. A dotted eighth used to be the really hip delay, but now there's a lot of swing going on, so it's better just to stay with the quarter and an eighth. Even the eighth sometimes doesn't work if the swing's really heavy. Sometime you have to fish around somewhere between triplets, but I always have a trusty calculator around or usually studios have a delay chart.

*Do you have an approach to using EQ?*
I have different approaches. I think most engineers probably do. If I'm recording vocals, I like to roll off quite a bit on the bottom end so the compressor doesn't start kicking in and bringing up any low end rumble or noise. If I'm EQing a piano or something that's already been recorded, I sometimes roll off a lot of the bottom so I leave a lot of room for the bass and the kick drum to occupy. A lot of times I don't need anything under probably 100Hz. I'll do some rolling off with the filters and then I may take a bell curve and zone in on a couple of other woofy areas on certain instruments.

If the vocal is really harsh and I want to soften it up a bit, I may actually use the compressor side-chain on the SSL. The elements that I want to take out of the vocal I'll accentuate first on the EQ and then I'll flip it over to the side-chain function on the SSL so it's kinda like stepping down on that section when the vocals get edgy.

*Do you have a method for setting levels?*
I would say for most of the Dance stuff, I use the drums to just find out where the kick drum is. I would say my kick drum's probably around –3, –2. If you were to look at some of the faders, it's really wild. I would say the kick drum could be like up at 0 and the snare could be down at almost –20.

*How long does it take you to do a mix generally?*

I would say for one of these club records an 18-hour period would be about average. It takes me maybe four or five hours to get sounds and about five or six hours for the producer to arrange a ten-minute long 12" mix. If the song is only five or six minutes on the 24-track, then I'll be looking at an intro piece that we have to build, a main body, and an edit piece. So there's three mixes right there that have to be looked at. Then there's a radio version, a dub version, and an alternative mix. Quite often I'll do my vocal rides on the 12" club mix and the radio mix will be based off of the body of the club mix, so I'll be able to at least join the fader levels, if not fader mutes, from one mix to another. But there may be some subtle changes in the arrangement and that vocal level may not work so well in the radio mix, or a new element or two is brought in or out, which means you have to update anyway. So there's a lot of listening that I have to do.

*How do you approach panning?*

If I'm doing a Dance club record, I don't go extremely wide with what I consider important elements, which would be kick, snares, hi- hats and cymbals. Because of the venues where the song is being played, if you pan a pretty important element on the left side, half the dance floor's not hearing it. So important elements like that I usually keep either up the middle or maybe like at 10:30 and 1:30. Lead vocals are almost always up the middle. Backgrounds I will pan pretty hard left and right. Congas and shakers I don't pan extremely hard but I will pan them out from the center. A lot of times I get these live recordings with congas that are in stereo. When you listen to it, it's got a low conga all the way to your left and a high conga all the way to your right. That's really nice for certain records, but when the tracks start getting really busy, it ruins the focus that your ear wants to have. If these elements are going to be used throughout the whole track and it's really busy, then I'm going to tighten up the image on that quite a bit.

*Do you have any listening tricks?*

I use an Auratone up the middle in mono. I'll do my basics on the Auratone and move to the Yamahas, but on the Yamahas I'll listen at a low level like 2 or 3. Then I'll graduate to 6 or 7, slide my chair back from the board, and just try to get more of an out-of-the-image listen. Then I'll listen up top on the big speakers really loud, just to make sure that I've got the bottom right. I have one particular room that I mix a lot at, Studio B at Quad Studios, I usually lay down with my head on the armrest of the couch in the back wall when I'm checking my bottom in that room. When I turn the big speakers up to like 8, it has to hit me a certain way. If it's not, then I know something's not right. Otherwise, I usually like to mix at a pretty relatively low volume for as long as I can.

When I'm doing my vocal line, I usually start low on the Auratone, and then I'll graduate to the Yamahas, especially for backgrounds. It's really important for me on these club records that the lead vocal is like biting your head off. We have 813s in the B room at Quad and if you can make them smooth up there, then they'll probably sound pretty smooth anywhere.

*There are a lot of effects tricks used in dance music. Give me an example of some of the things that you use.*
One of the things that I quite often do with the drums is use the small monitor faders as an audio subgroup. I'll get a little mix of the drums separate from the drum mix on the automated faders and bus them to one automated fader. Then I'll float that fader into an effects box, quite often the H3000. That way you can actually mute all the drums from the mix and then open that one fader and have all the drums going into one effect. You get this rhythmic pulsing, using either a flange or a filter pan delay or digital wah kind of effect, that creates an interesting thing that the remixer can just drop into with maybe just the kick drum underneath to keep the bottom end pumping. That's one of my popular effects.

I'll do the same thing with backgrounds or with the lead on another completely different fader. You can mute the lead vocal from the mix and then slide this other fader up and then the lead vocal has a real long delay like a quarter note.

I also like these things called "kick booms" where you take the kick drum and send it into a large reverb so it explodes. To a certain degree, I'm always playing. I'm always trying to reinvent the wheel, I guess.

# Bruce Swedien

P erhaps no one else in the studio world can so aptly claim the moniker of "Godfather of Recording" as Bruce Swedien. Universally revered by his peers, Bruce has earned that respect thanks to years of stellar recordings for the cream of the musical crop. His credits could fill a book alone, but legends like Count Basie, Lionel Hampton, Stan Kenton, Duke Ellington, Woody Herman, Oscar Peterson, Nat "King" Cole, George Benson, Mick Jagger, Paul McCartney, Edgar Winter and Jackie Wilson are good places to start. Then comes Bruce's Grammy winning projects which include Michael Jackson's *Thriller* (the biggest selling record of all time), *Bad* and *Dangerous,* and Quincy Jones' *Back on the Block* and *Juke Joint.* As one who has participated in the evolution of modern recording from virtually the beginning as well as being one of its true innovators, Bruce is able to give insights on mixing from a perspective that few of us will ever have.

*Do you have a philosophy about mixing that you follow?*
The only thing I could say about that is everything that I do in music, mixing or recording or producing, is music driven. It comes from my early days in the studio with Duke Ellington and from there to Quincy. I think the key word in that philosophy is what I would prefer to call responsibility. From Quincy — no one's influenced me more strongly than Quincy — I've learned that when we go into the studio our first thought should be that our responsibility is to the musical statement that we're going to make and to the individuals involved. And I guess that's really the philosophy that I follow.

*Responsibility in that you want to present the music in its best light?*
To do it the best way that I possibly can. To use everything at my disposal to not necessarily recreate an unaltered acoustic event, but to present either my concept of the music or the artist's concept of the music in the best way that I can.

*Is your concept ever opposed to the artist's concept?*
It's funny but I don't ever remember running into a situation where there's been a conflict. Maybe my concept of the sonics of the music might differ at first with the artist, but I don't ever remember it being a serious conflict.

*I would think that you're hired because of your overall concept.*
I have a feeling that's true, but I'm not really sure. I think probably my range of musical background helps a lot in that I studied piano for eight years and as a kid I spent a lot of time listening to classical music. So when it comes to depth of musical experience, I think that's one reason that people will turn to me for a project.

*Do you think that starting out without the benefit of the vast amount of technology that we have today has helped you?*
Oh, definitely. Absolutely. No question. And I think what's helped me more is that I was the right guy in the right place at the right time at Universal in Chicago. Bill Putnam, who was my mentor and brought me from Minneapolis as a kid, saw or heard something in me that I guess inspired some confidence. From there I got to work with people like Duke Ellington, Count Basie, Woody Herman, Stan Kenton, Oscar Peterson and so on. One of the thrilling parts about the late 50's at Universal in Chicago was that I literally learned microphone technique with Count Basie and Duke Ellington, and these guys were in love with the recording process.

*Really? I was under the impression they only recorded because they had to.*
No. Absolutely not. Now there were some band leaders that were that way, although I can't think of anybody offhand, but most of them just loved being there. The guy that I think was most formative in my early years as a kid was probably Count Basie. I did a lot of records with that band.

*How were you influenced?*
I came into the industry at that level as a real youngster. In 1958 I was only 20 years old and I started right out working with Stan Kenton, and a couple of years later Count Basie, Duke Ellington, Quincy and so on. But I was not in love with the status quo that was part of the recording industry at the time. The goal of music recording in the late 50's was to present the listener with a virtually unaltered acoustic event and that wasn't terribly exciting to me. I loved it, but I wanted my imagination to be part of the recording.

Another guy who bumped into that who I didn't work with but I got to meet in the early 60's at Universal was Les Paul. There was one record that I remember that came out when I was in high school in 1951 that changed popular music forever and it was Les Paul and Mary Ford's "How High the Moon," which was an absolutely incredible thing. I couldn't wait to get to the record store to buy it so I could try to figure out what that was all about. At that point in time, I think a whole segment of the record buying public made a left turn in that the records of the day were pretty much, as I said, an unaltered acoustic event and we were trying to put the listener in the best seat in the house. But all of a sudden this record came along without a shred of reality in it and a whole segment of the record buying public said, "This is what we want."

*That being said, can you hear that sonic space in your head before you start to mix?*
No. That's the wonderful part about it.

*Is your approach to mixing each song generally the same then?*
I'll take that a step further and I'll say it's never the same, and I think I have a very unique imagination. I also have another problem in that I hear sounds as colors in my mind. Frequently when I'm EQing or checking the spectrum of a mix or a piece of music, if I don't see the right colors in it I know the balance is not there.

*Wow! Can you elaborate on that?*
Well, low frequencies appear to my mind's eye as dark colors, black or brown, and high frequencies are brighter colors. Extremely high frequencies are gold and silver. It's funny, but that can be very distracting. It drives me crazy sometimes. There is a term for it but I don't know what it's called.

*What are you trying to do then, build a rainbow?*
No, it's just that if I don't experience those colors when I listen to a mix that I'm working on, I know that there's either an element missing or that the mix values aren't satisfying.

*How do you know what proportion of what color should be there?*
That's instinctive. Quincy has the same problem. It's terrible! Drives me nuts! But it's not a quantitative thing. It's just that if I focus on a part of the spectrum in a mix and don't see the right colors, it bothers me. I have a feeling it's a disease, but people have told me it isn't.

*How do you go about getting a balance? Do you have a method?*
No, it's purely instinctive. Another thing that I've learned from Quincy, but started with my work with Duke Ellington, is to do my mixing reactively not cerebrally. When automated mixing came along, I got really excited because I thought, "At last, here's a way for me to preserve my first instinctive reaction to the music and the mix values that are there." You know how frequently we'll work and work and work on a piece of music and we think, "Oh boy, this is great. Wouldn't it be great if it had a little more of this or a little more of that." Then you listen to it in the cold gray light of dawn and it sounds like shit. Well, that's when the cerebral part of our mind takes over, pushing the reactive part to the background, so the music suffers.

*Do you start to do your mix from the very first day of tracking?*
Yes, but again I don't think that you can say any of these thoughts are across the board. There are certain types of music that grow in the studio. You go in and start a rhythm track and think you're gonna have one thing and all of a sudden it does a sharp left and it ends up being something else. While there are other types of music where I start the mix even before the musicians come to the studio. I'll give you a good example of something. On Michael's *History* album, for the song "Smile, Charlie Chaplain," I knew what that mix would be like two weeks before the musicians hit the studio.

*From listening to the demo?*
No. It had nothing to do with anything except what was going on in my mind because Jeremy Lubbock, the orchestra arranger and conductor, and I had talked about that piece of music and the orchestra that we were going to use. I came up with a studio setup that I had used with the strings of the Chicago Symphony many years before at Universal where the first violins are set up to the left of the conductor and the second violins to the right, the violas behind the first fiddles and the cello behind the second fiddles, which is a little unusual. So I had that whole mix firmly in mind long before we did it.

*So sometimes you do hear the final mix before you start.*
Sometimes, but that's rare.

*Where do you generally build your mix from?*
It's totally dependent on the music. Always. But if there was a method of my approach, I would say the rhythm section. You usually try to find the motor and then build the car around it.

*Some people say they always put the bass up first, some from the snare, some the overheads…*

No, I don't think I have any set way. I think it would spoil the music to think about it that much.

*I guess you don't have any kind of method for setting balances.*

Starting the bass at –5 or something? Boy, that would be terrible. I couldn't do that if my life depended on it.

*Do you have a method for panning?*

I don't think I have any approach to it. I generally do whatever works with the music that I'm doing.

*So it's just something that hits you when you're doing it?*

Yeah, that's really the way it works. It'll be an idea; whether it's panning or a mix value or an effect or whatever, and I'll say, "Ooh, that's great. I'm gonna do that."

*What level do you usually monitor at?*

That's one area where I think I've relegated it to a science. For the near-field speakers, I use Westlake BBSM8s and I try not to exceed 85dB SPL. On the Auratones I try not to exceed 83. What I've found in the past few years, I use the big speakers less and less with every project.

*Are you listening in mono on the Auratones?*

Stereo.

*Do you listen in mono much?*

Once in awhile. I always check it because there's some places where mono is still used.

*I love the way you sonically layer things when you mix. How do you go about getting that?*

I have no idea. If I knew, I probably couldn't do it as well. It's purely reactive and instinctive. I don't have a plan. Actually, what I will do frequently when we're layering with synths and so on, is to add some acoustics to the synth sounds. I think this helps in the layering in that the virtual direct sound of most synthesizers is not too interesting, so I'll send the sound out to the studio and use a coincident pair of mics to blend a little bit of acoustics back with the direct sound. Of course, it adds early reflections to the sound, which reverb devices can't do. That's the space before the onset of reverb where those early reflections occur.

*So what you're looking for more than anything is early reflections?*
I think that's a much overlooked part of sound because there are no reverb devices that can generate that. It's very important. Early reflections will usually occur under 40 milliseconds. It's a fascinating part of sound.

*When you're adding effects, are you using mostly reverbs or delays?*
A combination. Lately though, I have been kinda going through a phase of using less reverb. I've got two seven foot high racks full of everything. I have an EMT250, a 252, and all the usual stuff. All of it I bought new. No one else has ever used them. It's all in pretty good shape too.

*Do you have any listening tricks?*
You know what? Since I moved from California (I live in Connecticut now and I'm not going back), one of the things that I miss is my time in the car. I had a Ford Bronco with an incredible sound system and I still kinda miss that great listening environment.

*Do you do all your work at your facility now?*
No, wherever they'll have me. I love it here, but my studio's dinky. I have an older little 40-input Harrison and a 24-track. The Harrison is a wonderful desk. It's a 32 series and the same as the one I did *Thriller* on. Actually I think that's one of the most underrated desks in the industry. It's all spiffed up with a beautiful computer and Neve summing amps. It's just fabulous.

*Didn't you used to have a couple of Neves put together?*
I did have a beautiful Neve but after I finished Michael's *History* album and Quincy's *Juke Joint,* I was kind of burned out and very, very tired, so I told my wife as we were having breakfast one morning, "Honey, I'm gonna get rid of this damn studio at home and I don't ever want to have another at home." Six months later I was buying a console. I guess once a junkie, always a junkie.

*How long does it usually take you to do a mix?*
That can vary. I like to try not to do more than one song a day unless it's a real simple project, and then I like to sleep on a mix and keep it on the desk overnight. That's one of the advantages of having my little studio at home.

*I know that a lot of your projects are really extensive in terms of tracks.*
But that's not so much true any more. I start a mix tomorrow here at home for EMI in Portugal of a Portuguese band. It's all on one 24-track tape.

*How many versions of a mix do you do?*
Usually one. Although when I did "Billie Jean," I did 91 mixes of that thing and the mix that we finally ended up using was mix 2. I had a pile of 1/2" tapes to the ceiling. And we thought, "Oh man, it's getting better and better." [*laughs*]

*What are you using for a mastering machine these days?*
I have an Ampex ATR with both 1/2" and 1/4" heads. I also have a Mitsubishi 86HS that I don't really use any more. The ATR is my favorite. I bought it new and nobody else has ever used it.

*So it's mostly 1/2"?*
No. 1/4" is wonderful and I use it a lot. 1/4" has a little different sound. It's a little more mellow. 1/2", because of the tape width, has phenomenal transient response. If you're doing R&B or Rock or Pop music, then that's a great choice. But this band from Portugal that I'm mixing is Fado music and it's very somber and pretty and soft, so I'm gonna probably do that on 1/4".

*I haven't heard anybody mention 1/4" in a long time.*
It's typical with pretty music and I think it's the better format. It's very, very lovely.

*Do you have an approach to using EQ?*
I don't think I have a philosophy about it. What I hate to see is an engineer or producer start EQing before they've heard the sound source. To me it's kinda like salting and peppering your food before you've tasted it. I always like to listen to the sound source first, whether it be on tape or live, and see how well it holds up without any EQ or whatever.

*That being the case, do you have to approach things differently if you're just coming in to do the mix?*
Not usually. But I'm not really crazy about listening to other people's tapes, I gotta tell you that. But I consider myself fortunate to be working, so that's the bottom line [*laughs*].

*Do you add effects as you go?*
There's probably only two effects that I use on almost everything and that's the EMT 250 and the 252. I love those reverbs. There's nothing in the industry that comes close to a 250 or a 252.

*What are you using the 252 on?*
I love the 252 on vocals with the 250 program. It's close to a 250, but it's kinda like a 250 after taxes. It's wonderful, but there's nothing like a 250.

*What do you do to make a mix special?*
I wish I knew. I have no idea. But the best illustration of something special is when we were doing "Billie Jean" and Quincy said, "Okay, this song has to have the most incredible drum sound that anybody has ever done but it also has to have one element that's different, and that's sonic personality." So I lost a lot of sleep over that. What I ended up doing was building a drum platform and designing some special little things like a bass drum cover and a flat piece of wood that goes between the snare and the hi-hat. And the bottom line is that there aren't many pieces of music where you can hear the first three or four notes of the drums and immediately tell what piece of music it is. But I think that is the case with "Billie Jean," and that I attribute to sonic personality. But I lost a lot of sleep over that one before it was accomplished.

*Do you determine that personality before you start to record?*
Not really. But in that case I got to think about the recording setup in advance. And of course, I have quite a microphone collection that goes with me everywhere (17 Anvil cases!) and that helps a little bit in that they're not beat up.

*Are most of the projects that you do these days both tracking and mixing?*
I don't know what's happened but I don't get called to record stuff very much these days. People are driving me nuts with mixing and I love it, but I kinda miss tracking. A lot of people think that since I moved to Connecticut I retired or something, but that's the last thing I'd want to do. You know what Quincy and I say about retiring? Retiring is when you can travel around and get to do what you want. Well, I've been doing that all my life. I love what I do and I'm just happy to be working. So that's the bottom line.

# CHAPTER 33

# John X

Yes, his effects are loaded and he's not afraid to use them. John X. Volaitis is one of the new breed of engineers who's thrown off his old school chains and ventured into the world of remixes (known to some as Techno, Trance, Industrial, Ambient, or any one of about ten other different names). Along with his partner Danny Saber, John has done recent re-mixes for such legends as David Bowie (*Dead Man Walking,* "Little Wonder") and U2 ("Staring at the Sun"), as well as Marilyn Manson (*Horrible People*), Garbage ("Stupid Girl") and a host of others. As you'll see, the X man's methods are both unique and fun.

*When people send you tapes for remixes, what are they actually sending you?*
Far too often they send us the entire multi-track when all we need is one track of that. Usually you could just send the lead vocal, time code and a start time and tempo because half the time we'll change the key and the tempo anyway. David Bowie's "Little Wonder," for instance, we did in almost half time. That means you're pretty much throwing out most of the original tracks because you can't use any of that stuff to begin with.

Sometimes that's inappropriate though. We just did one for U2, "Staring at the Sun," which they're really happy with. Part of why I think they're happy is the fact that we didn't butcher them at all. We kept a little piece of everybody because they're a band. The one thing you learn is that when you remix for a band, you can't have the singer and the guitar player in the track but not have the bass player and the drummer in it because it creates total warfare for them that's gonna make them say, "Look, let's not use that." So you find one little thing, like some thick fill that the drummer did or the bass player making some noise at the beginning of the song, and use that. Maybe it's the only thing that you can really loop and get into the track and make it dancey, but it's something that's gonna let them say, "Hey man, that's me." As long as they know they're in there, they're fine.

One special thing I've noticed about the remixes is that if you don't have a really great vocal performance to begin with, you're screwed. The same old rule still applies. If you got a good performance on somebody, you can almost do anything to it and it's still good. With somebody like Bowie or Bono, those guys are *the* cats, so it sounds great right off.

*Where do you build your mix from?*
I generally have to start with the loops. You've got to find the main loop or the combination of loops that creates the main groove. Sometimes the loops may have a lot of individual drums, but they're usually not crucial rhythmic elements. They can be accents and they can be stuff that just pops up in a break here and there.

*Do you use a lot of compression?*
I use it a lot. Not always in great amounts but I tend to try to get some handle on the peaks. Loops I rarely mess with. If somebody's got a loop and a certain groove that they like, I almost always leave those things alone because they start getting real squirrelly if you mess with them. All of a sudden the groove can change radically. Anything else, I don't mind slammin' the hell out of as long as it sounds the way I want it to sound. I don't even have a rule about it.

*What's your approach to panning?*
I try not to waste the sides on anything that's not really, really actively stereo like a lot of stereo patches that come out of people's keyboards. To me, most of those are boring. They're not really doing anything and they're a waste of the sides so I'll just tuck them up a little bit. I'll have them left and right but not always hard. What I always try to do is keep my effects and delays or my radical panning stuff hard left and right, which makes it feel like the more radical stuff is projecting a little bit further than the rest of the band. It just gives a different sort of depth perception.

*What about adding effects? Do you add them as you go along or do you get a balance and then add them?*
I do both. I love effects. For years I acted as a total purist. "I'm gonna bypass everything in the room and go to tape." That's really cool for some stuff but what it was doing was getting me into a mindset where I couldn't even think about putting effects on stuff with any real imagination. All I was coming up with was sort of the same thing you always hear on conservative albums.

Now as far as I'm concerned I'll go haywire with that stuff. I'll record effects on tracks and if I don't like them, I'll just erase them. It's no big deal.

That sounds chaotic, but usually I'm trying to accomplish something very special. For example, sometimes I'll mult the lead vocal to a few different channels. Then each will be EQed completely differently, some compressed, some de-essed severely and sent to long reverbs, and some super filtered and sent to some weird pitch modulation stuff. Then I'll flip around from syllable to syllable throughout the verses and have the entire effect structured around the voice constantly shifting, yet leave something consistent from the main vocal track. As far as the amount of dry vocal, that's arbitrary. Sometimes, you find some spots where you feel like it's right and you leave it there and then just sort of tune in that other stuff so it becomes something natural. But sometimes it's something special to where anyone who listens to it can't even figure out what the hell is on there. You don't have enough time to identify the flanger because by the time you think it might be a flanger it's already turned into three other effects. You have to do it in such a way so it's not a distraction and it's not defeating the lead vocalist, though.

*Do you have a method for setting up your effects?*
Yeah, I go right for the Eventide stuff first. I'm definitely a snob that way. I go for the 4000, the 3500, and the 3000. Then I'm gonna go into the 480s and stuff, depending on what the people have in them. I actually like the 224X with the Resonant Chords. I love that patch and use that thing a lot, and I use the PCM70 with their "Rhyme in C-minor BPM, Rhyme in C-major BPM" patch a lot also. You can generate pitches with a nice resonance but be able to tune the pitches to the key of your track and set the delay time almost instantly. I've taken some really dud parts and just really brought them to life with those. I've been making loops with them too, like tuned drum loops that are really usable. I definitely go for the weird stuff first. I only put up maybe one reverb. Reverb has a way of piling up underneath your track, so there's a lot more of it underneath the track than there is on top of it. I'll usually try to keep the amount of reverb down to one special item for a distance perspective kind of thing, just to let you know that someone's back there, but not so much to drown the band. I prefer the shorter, weirder stuff, definitely.

*What do you bring to the studio with you?*
I have a bunch of mostly funky gear. A lot of stuff I bring is not engineer related, its more music related. I'll bring some weird toys like an ancient AKAI sampler that has knobs on the front that lets me do all kinds of weird stuff. I have some weird vocal processors like the Digitech Vocalist VH5, which I use quite often, and a Korg version that was made years ago. It's a really super cheesy vocoder with some weird key bending stuff that you can do. Other than that, I bring a Lexicon Vortex and a bunch of weird little pedals. That's about it. I like to bring the stuff you know you're not gonna find in a control room.

*What format are you mixing to?*
For remixes, usually DAT with some nice converters. If it's something that I really think has come out special I'll put it on 1/2", but usually I don't bother with 1/2" on remixes because we'd send them off and the people would master off the DATs anyway. It's kind of a waste of tape and time to even bother sending the 1/2" if the people aren't gonna use them. A lot of independent labels aren't even mastering the mixes either. They're transferring the DATs straight across and that's it. I just sort of became aware of this in the last few months.

*Since you're building up the track from scratch, how long does it take you on a remix?*
Usually we try to do them in two days. My partner, Danny Saber, will go in the first day and he'll do his musical arrangement, such as playing all the parts and laying out the loops and all that stuff. The second day is mine and that's it. Sometimes it slips into three, depending on how elaborate I get on that second day.

When I'm mixing you don't really get the impression that people are working in there. The vibe is whatever it is because that's my day, it's my show and I can do what I want with it. I rarely stress out about anything and it's always gonna be complete mayhem and chaos. My main assistant has to be wearing a lab coat that says "Patch Boy" on it and I have my own dark blue one. We've found that giving the assistants those lab coats gives them a really new sense of importance. At first we were joking about it but now it's like, "Damn, look at these guys! They have become serious." If you ask them a question they're right on it [*laughing*]. And the best thing is when clients come in who've never worked with us, the assistants could tell them anything and they believe them because they're wearing that lab coat. It's like "The doctor just told me that this is the way it's gonna go down so I believe them." The whole thing is really fun!

# Glossary

**Attack**  The first part of a sound. On a compressor/limiter, a control that affects how that device will respond to the attack of a sound.

**Attenuation**  A decrease in level.

**Automation**  A system that memorizes, then plays back the position of all faders and mutes on a console.

**Bandwidth**  The number of frequencies that a device will pass before the signal degrades. A human being can supposedly hear from 20Hz to 20kHz, so the bandwidth of the human ear is 20 to 20kHz.

**Bass Management**  A circuit which utilizes the subwoofer in a 5.1 system to provide bass extension for the five main speakers. The Bass Manager steers all frequencies below 80Hz into the subwoofer along with the LFE (see **LFE**) source signal.

**Bass Redirection**  Another term for Bass Management.

**Bit Rate**  The transmission rate of a digital system.

**Bit Splitter**  In order to record a signal with a 20-bit word length onto a recorder that is only 16-bit, the digital word is "split" across two tracks instead of one.

**Buss**  A signal pathway.

**Chamber (Reverb)**  A method to create artificial reverberation using a tiled room with a speaker and several microphones placed in the room.

**Chorus**  A type of signal processor where a de-tuned copy is mixed with the original signal which creates a fatter sound.

**Comb Filter**  A distortion produced by combining an electronic or acoustic signal with a delayed copy of itself. The result is peaks and dips introduced into the frequency response. This is what happens when a signal is flanged (see **Flanging**).

**Cut Pass** A playback of the song in which the engineer programs the mutes only into the automation computer in order to clean up the mix.

**Cut** To decrease, attenuate or make less.

**Data Compression** Takes multiple data streams (as in 6-channel surround sound) and compresses them into a single data stream for more efficient storage and transmission. Supposedly some of what is normally recorded before compression is imperceptible, with the louder sounds masking the softer ones. As a result, some of this data can be eliminated since it's not heard anyway. This selective approach, determined by psychoacoustic research, is the basis for "lossy" compression. It is debatable however, how much data can actually be thrown away (or compressed) without an audible sacrifice. Dolby AC-3 and DTS are both lossy compression schemes.

**DAW** A Digital Audio Workstation. A computer with the appropriate hardware and software needed to digitize and edit audio.

**DDL** Digital Delay Line. Same as a digital delay processor.

**Decay** The time it takes for a signal to fall below audibility.

**Delay** A type of signal processor that produces distinct repeats (echoes) of a signal.

**Dolby Digital®** A data compression method, otherwise known as AC-3, which uses psychoacoustic principles to reduce the number of bits required to represent the signal. Bit rates for 5.1 channels range from 320kbps for sound on film to 384kbps for digital television and up to 448kbps for audio use on DVD. AC-3 is also what's known as a "lossy" compressor (see **Lossy Compression**) that relies on psychoacoustic modeling of frequency and temporal masking effects to reduce bits by eliminating those parts of the signal thought to be inaudible. The bit rate reduction achieved at a nominal 384kbps is about 10:1.

**Down-mix** To automatically extract a stereo or mono mix from an encoded surround mix.

**DTS** A data compression method developed by Digital Theater Systems using waveform coding techniques that takes six channels of audio (5.1) and folds them into a single digital bit stream. This differs from Dolby Digital® in that the data rate is a somewhat higher 1.4Mbs, which represents a compression ratio of about 4:1. DTS is also what's known as a "lossy" compression (see **Lossy Compression**).

**DTV** Digital Television.

**Element** A component or ingredient of the sound or groove.

**Elliptical EQ** A special equalizer built especially for vinyl disc mastering that takes excessive bass energy from either side of a stereo signal and directs it to the center. This was to prevent excessive low frequency energy from cutting through the groove wall and destroying the master lacquer.

**Equalizer** A tone control that can vary in sophistication from very simple to very complex (see **Parametric Equalizer**).

**Exciter** An outboard effects device that uses phase manipulation and harmonic distortion to produce high frequency enhancement of a signal.

**5.1** A speaker system that uses three speakers across the front and two stereo speakers in the rear, along with a subwoofer.

**Flanging** The process of mixing a copy of the signal back with itself, but gradually and randomly slowing the copy down to cause the sound to "whoosh" as if it were in a wind tunnel. This was originally done by holding a finger against a tape flange (the metal part that holds the tape on the reel), hence the name.

**Fletcher-Munson Curves** A set of measurements that describes how the frequency response of the ear changes at different sound pressure levels. For instance, we generally hear very high and very low frequencies much better as the overall sound pressure level is increased.

**Groove** The pulse of the song and how the instruments dynamically breathe with it.

**HDCD** High-Definition Compatible Digital® is a process which encodes 20 bits of information onto a standard 16-bit CD, while still remaining compatible with normal CD players.

**LFE**   Low Frequency Effects channel. This is a special channel of 5Hz to 120Hz information primarily intended for special effects such as explosions in movies. The LFE has an additional 10dB of headroom in order to accommodate the required level.

**Make-up Gain**   A control on a compressor/limiter that applies additional gain to the signal. This is required since the signal is automatically decreased when the compressor is working. Make-up Gain "makes up" the gain and brings it back to where it was prior to being compressed.

**MDM**   Modular Digital Multitrack. A low cost 8-track digital recorder that can be grouped together to configure as many tracks as are needed. The Tascam DA-88 and Alesis ADAT are the most popular MDMs.

**MLP**   Meridian Lossless Packing. This is a data compression technique designed specifically for high quality (96kHz/24bit) sonic data. MLP differs from other data compression techniques in that no significant data is thrown away, thereby claiming the "Lossless" moniker. MLP is also a standard for the 96kHz/24bit portion of the new DVD-Audio disc and is licensed by Dolby Labs.

**MO**   Magneto Optical. A re-writeable method of digital storage utilizing an optical disc. Each disc stores from 250MB to 4.3GB and may be double-sided. Its widespread use has been limited by its slow access time.

**Modulate**   The process of adding a control voltage to a signal source in order to change its character. For example, modulating a short slap delay with a .5Hz signal will produce chorusing (see **Chorus**).

**Parametric Equalizer**   A tone control where the gain, frequency and bandwidth are all variable.

**Mute**   An On/Off switch. To mute something would mean to turn it off.

**Phantom Image**   In a stereo system, if the signal is of equal strength in the left and right channels, the resultant sound appears to come from in between them. This is a phantom image.

**Phase Shift** The process during which some frequencies (usually those below 100Hz) are slowed down ever so slightly as they pass through a device. This is usually exaggerated by excessive use of equalization and is highly undesirable.

**Plate (Reverb)** A method to create artificial reverberation using a large steel plate with a speaker and several transducers connected to it.

**Pre-delay** A variable length of time before the onset of reverberation. Pre-delay is often used to separate the source from the reverberation so the source can be heard more clearly.

**Pultec** An equalizer sold during the 50's and 60's by Western Electric that is highly prized today for its smooth sound.

**Q** Bandwidth of a filter or equalizer.

**Ratio** A control on a compressor/limiter that determines how much compression or limiting will occur when the signal exceeds threshold.

**Range** On a gate or expander, a control that adjusts the amount of attenuation that will occur to the signal when the gate is closed.

**Recall** A system that memorizes the position of all pots and switches on a console. The engineer must still physically reset the pots and switches back to their previous positions as indicated on a video monitor.

**Release** The last part of a sound. On a compressor/limiter, a control that affects how that device will respond to the release of a sound.

**Reverb** A type of signal processor that reproduces the spatial sound of an environment (e.g., the sound of a closet or locker room or inside an oil tanker).

**Punchy** A description for a quality of sound that infers good reproduction of dynamics with a strong impact. Sometimes means emphasis in the 200Hz and 5kHz areas.

**Return** Inputs on a recording console especially dedicated for effects devices such as reverbs and delays. The Return inputs are usually not as sophisticated as normal channel inputs on a console.

**SDDS** Sony Dynamic Digital Sound. Sony's digital delivery system for the cinema. This 7.1 system features five speakers across the front, stereo speakers on the sides, plus a subwoofer.

**Selsync** Short for Selective Synchronization. This is the process of using the record head on a tape machine to do simultaneous playback of previous recorded tracks while recording. This process is now called *overdubbing*.

**Sibilance** A rise in the frequency response in a vocal where there's an excessive amount of 5kHz, resulting in the "S" sounds being overemphasized.

**SMART Content** System Management Audio Resource Technique. This feature allows the producer to control the way the multi-channel audio is played back in stereo by saving one of 16 mixdown coefficients as control information to a data channel on the DVD-A.

**SPL** Sound Pressure Level.

**Synchronization** When two devices, usually storage devices such as tape machines, DAW's or sequencers, are locked together with respect to time.

**Sub** Short for subwoofer.

**Subwoofer** A low frequency speaker with a frequency response from about 25Hz to 120Hz.

**Tape Slap** A method to create a delay effect by using the repro head of a tape machine (which is after the record head).

**Threshold** The point at which an effect takes place. On a compressor/limiter for instance, the Threshold control adjusts the point at which compression will take place.

**Track Sharing** When a single track shares more than one instrument. For instance, when a percussion part is playing on a guitar solo track in places that the guitar has not been recorded.

**TV Mix** A mix without the vocals so the artist can sing live to the back tracks during a television appearance.

# Decay Timing Chart

| BPM | 1/2 Note | 1/4 Note | 1/8 Note | 1/16 Note | 1/4 Triplet | 1/8 Triplet | Dotted 1/4 | Dotted 1/8 |
|-----|----------|----------|----------|-----------|-------------|-------------|------------|------------|
| 60 | 1000.00 | 500.00 | 250.00 | 125.00 | 333.50 | 166.75 | 750.00 | 375.00 |
| 61 | 983.61 | 491.80 | 245.90 | 122.95 | 328.03 | 164.02 | 737.70 | 368.85 |
| 62 | 967.74 | 483.87 | 241.94 | 120.97 | 322.74 | 161.37 | 725.81 | 362.90 |
| 63 | 952.38 | 476.19 | 238.10 | 119.05 | 317.62 | 158.81 | 714.29 | 357.14 |
| 64 | 937.50 | 468.75 | 234.38 | 117.19 | 312.66 | 156.33 | 703.13 | 351.56 |
| 65 | 923.08 | 461.54 | 230.77 | 115.38 | 307.85 | 153.92 | 692.31 | 346.15 |
| 66 | 909.09 | 454.55 | 227.27 | 113.64 | 303.18 | 151.59 | 681.82 | 340.91 |
| 67 | 895.52 | 447.76 | 223.88 | 111.94 | 298.66 | 149.33 | 671.64 | 335.82 |
| 68 | 882.35 | 441.18 | 220.59 | 110.29 | 294.26 | 147.13 | 661.76 | 330.88 |
| 69 | 869.57 | 434.78 | 217.39 | 108.70 | 290.00 | 145.00 | 652.17 | 326.09 |
| 70 | 857.14 | 428.57 | 214.29 | 107.14 | 285.86 | 142.93 | 642.86 | 321.43 |
| 71 | 845.07 | 422.54 | 211.27 | 105.63 | 281.83 | 140.92 | 633.80 | 316.90 |
| 72 | 833.33 | 416.67 | 208.33 | 104.17 | 277.92 | 138.96 | 625.00 | 312.50 |
| 73 | 821.92 | 410.96 | 205.48 | 102.74 | 274.11 | 137.05 | 616.44 | 308.22 |
| 74 | 810.81 | 405.41 | 202.70 | 101.35 | 270.41 | 135.20 | 608.11 | 304.05 |
| 75 | 800.00 | 400.00 | 200.00 | 100.00 | 266.80 | 133.40 | 600.00 | 300.00 |
| 76 | 789.47 | 394.74 | 197.37 | 98.68 | 263.29 | 131.64 | 592.11 | 296.05 |
| 77 | 779.22 | 389.61 | 194.81 | 97.40 | 259.87 | 129.94 | 584.42 | 292.21 |
| 78 | 769.23 | 384.62 | 192.31 | 96.15 | 256.54 | 128.27 | 576.92 | 288.46 |
| 79 | 759.49 | 379.75 | 189.87 | 94.94 | 253.29 | 126.65 | 569.62 | 284.81 |
| 80 | 750.00 | 375.00 | 187.50 | 93.75 | 250.13 | 125.06 | 562.50 | 281.25 |
| 81 | 740.74 | 370.37 | 185.19 | 92.59 | 247.04 | 123.52 | 555.56 | 277.78 |
| 82 | 731.71 | 365.85 | 182.93 | 91.46 | 244.02 | 122.01 | 548.78 | 274.39 |
| 83 | 722.89 | 361.45 | 180.72 | 90.36 | 241.08 | 120.54 | 542.17 | 271.08 |
| 84 | 714.29 | 357.14 | 178.57 | 89.29 | 238.21 | 119.11 | 535.71 | 267.86 |
| 85 | 705.88 | 352.94 | 176.47 | 88.24 | 235.41 | 117.71 | 529.41 | 264.71 |
| 86 | 697.67 | 348.84 | 174.42 | 87.21 | 232.67 | 116.34 | 523.26 | 261.63 |
| 87 | 689.66 | 344.83 | 172.41 | 86.21 | 230.00 | 115.00 | 517.24 | 258.62 |
| 88 | 681.82 | 340.91 | 170.45 | 85.23 | 227.39 | 113.69 | 511.36 | 255.68 |
| 89 | 674.16 | 337.08 | 168.54 | 84.27 | 224.83 | 112.42 | 505.62 | 252.81 |
| 90 | 666.67 | 333.33 | 166.67 | 83.33 | 222.33 | 111.17 | 500.00 | 250.00 |
| 91 | 659.34 | 329.67 | 164.84 | 82.42 | 219.89 | 109.95 | 494.51 | 247.25 |
| 92 | 652.17 | 326.09 | 163.04 | 81.52 | 217.50 | 108.75 | 489.13 | 244.57 |
| 93 | 645.16 | 322.58 | 161.29 | 80.65 | 215.16 | 107.58 | 483.87 | 241.94 |
| 94 | 638.30 | 319.15 | 159.57 | 79.79 | 212.87 | 106.44 | 478.72 | 239.36 |
| 95 | 631.58 | 315.79 | 157.89 | 78.95 | 210.63 | 105.32 | 473.68 | 236.84 |
| 96 | 625.00 | 312.50 | 156.25 | 78.13 | 208.44 | 104.22 | 468.75 | 234.38 |
| 97 | 618.56 | 309.28 | 154.64 | 77.32 | 206.29 | 103.14 | 463.92 | 231.96 |
| 98 | 612.24 | 306.12 | 153.06 | 76.53 | 204.18 | 102.09 | 459.18 | 229.59 |
| 99 | 606.06 | 303.03 | 151.52 | 75.76 | 202.12 | 101.06 | 454.55 | 227.27 |
| 100 | 600.00 | 300.00 | 150.00 | 75.00 | 200.10 | 100.05 | 450.00 | 225.00 |
| 101 | 594.06 | 297.03 | 148.51 | 74.26 | 198.12 | 99.06 | 445.54 | 222.77 |
| 102 | 588.24 | 294.12 | 147.06 | 73.53 | 196.18 | 98.09 | 441.18 | 220.59 |
| 103 | 582.52 | 291.26 | 145.63 | 72.82 | 194.27 | 97.14 | 436.89 | 218.45 |
| 104 | 576.92 | 288.46 | 144.23 | 72.12 | 192.40 | 96.20 | 432.69 | 216.35 |
| 105 | 571.43 | 285.71 | 142.86 | 71.43 | 190.57 | 95.29 | 428.57 | 214.29 |
| 106 | 566.04 | 283.02 | 141.51 | 70.75 | 188.77 | 94.39 | 424.53 | 212.26 |
| 107 | 560.75 | 280.37 | 140.19 | 70.09 | 187.01 | 93.50 | 420.56 | 210.28 |

| BPM | 1/2 Note | 1/4 Note | 1/8 Note | 1/16 Note | 1/4 Triplet | 1/8 Triplet | Dotted 1/4 | Dotted 1/8 |
|---|---|---|---|---|---|---|---|---|
| 108 | 555.56 | 277.78 | 138.89 | 69.44 | 185.28 | 92.64 | 416.67 | 208.33 |
| 109 | 550.46 | 275.23 | 137.61 | 68.81 | 183.58 | 91.79 | 412.84 | 206.42 |
| 110 | 545.45 | 272.73 | 136.36 | 68.18 | 181.91 | 90.95 | 409.09 | 204.55 |
| 111 | 540.54 | 270.27 | 135.14 | 67.57 | 180.27 | 90.14 | 405.41 | 202.70 |
| 112 | 535.71 | 267.86 | 133.93 | 66.96 | 178.66 | 89.33 | 401.79 | 200.89 |
| 113 | 530.97 | 265.49 | 132.74 | 66.37 | 177.08 | 88.54 | 398.23 | 199.12 |
| 114 | 526.32 | 263.16 | 131.58 | 65.79 | 175.53 | 87.76 | 394.74 | 197.37 |
| 115 | 521.74 | 260.87 | 130.43 | 65.22 | 174.00 | 87.00 | 391.30 | 195.65 |
| 116 | 517.24 | 258.62 | 129.31 | 64.66 | 172.50 | 86.25 | 387.93 | 193.97 |
| 117 | 512.82 | 256.41 | 128.21 | 64.10 | 171.03 | 85.51 | 384.62 | 192.31 |
| 118 | 508.47 | 254.24 | 127.12 | 63.56 | 169.58 | 84.79 | 381.36 | 190.68 |
| 119 | 504.20 | 252.10 | 126.05 | 63.03 | 168.15 | 84.08 | 378.15 | 189.08 |
| 120 | 500.00 | 250.00 | 125.00 | 62.50 | 166.75 | 83.38 | 375.00 | 187.50 |
| 121 | 495.87 | 247.93 | 123.97 | 61.98 | 165.37 | 82.69 | 371.90 | 185.95 |
| 122 | 491.80 | 245.90 | 122.95 | 61.48 | 164.02 | 82.01 | 368.85 | 184.43 |
| 123 | 487.80 | 243.90 | 121.95 | 60.98 | 162.68 | 81.34 | 365.85 | 182.93 |
| 124 | 483.87 | 241.94 | 120.97 | 60.48 | 161.37 | 80.69 | 362.90 | 181.45 |
| 125 | 480.00 | 240.00 | 120.00 | 60.00 | 160.08 | 80.04 | 360.00 | 180.00 |
| 126 | 476.19 | 238.10 | 119.05 | 59.52 | 158.81 | 79.40 | 357.14 | 178.57 |
| 127 | 472.44 | 236.22 | 118.11 | 59.06 | 157.56 | 78.78 | 354.33 | 177.17 |
| 128 | 468.75 | 234.38 | 117.19 | 58.59 | 156.33 | 78.16 | 351.56 | 175.78 |
| 129 | 465.12 | 232.56 | 116.28 | 58.14 | 155.12 | 77.56 | 348.84 | 174.42 |
| 130 | 461.54 | 230.77 | 115.38 | 57.69 | 153.92 | 76.96 | 346.15 | 173.08 |
| 131 | 458.02 | 229.01 | 114.50 | 57.25 | 152.75 | 76.37 | 343.51 | 171.76 |
| 132 | 454.55 | 227.27 | 113.64 | 56.82 | 151.59 | 75.80 | 340.91 | 170.45 |
| 133 | 451.13 | 225.56 | 112.78 | 56.39 | 150.45 | 75.23 | 338.35 | 169.17 |
| 134 | 447.76 | 223.88 | 111.94 | 55.97 | 149.33 | 74.66 | 335.82 | 167.91 |
| 135 | 444.44 | 222.22 | 111.11 | 55.56 | 148.22 | 74.11 | 333.33 | 166.67 |
| 136 | 441.18 | 220.59 | 110.29 | 55.15 | 147.13 | 73.57 | 330.88 | 165.44 |
| 137 | 437.96 | 218.98 | 109.49 | 54.74 | 146.06 | 73.03 | 328.47 | 164.23 |
| 138 | 434.78 | 217.39 | 108.70 | 54.35 | 145.00 | 72.50 | 326.09 | 163.04 |
| 139 | 431.65 | 215.83 | 107.91 | 53.96 | 143.96 | 71.98 | 323.74 | 161.87 |
| 140 | 428.57 | 214.29 | 107.14 | 53.57 | 142.93 | 71.46 | 321.43 | 160.71 |
| 141 | 425.53 | 212.77 | 106.38 | 53.19 | 141.91 | 70.96 | 319.15 | 159.57 |
| 142 | 422.54 | 211.27 | 105.63 | 52.82 | 140.92 | 70.46 | 316.90 | 158.45 |
| 143 | 419.58 | 209.79 | 104.90 | 52.45 | 139.93 | 69.97 | 314.69 | 157.34 |
| 144 | 416.67 | 208.33 | 104.17 | 52.08 | 138.96 | 69.48 | 312.50 | 156.25 |
| 145 | 413.79 | 206.90 | 103.45 | 51.72 | 138.00 | 69.00 | 310.34 | 155.17 |
| 146 | 410.96 | 205.48 | 102.74 | 51.37 | 137.05 | 68.53 | 308.22 | 154.11 |
| 147 | 408.16 | 204.08 | 102.04 | 51.02 | 136.12 | 68.06 | 306.12 | 153.06 |
| 148 | 405.41 | 202.70 | 101.35 | 50.68 | 135.20 | 67.60 | 304.05 | 152.03 |
| 149 | 402.68 | 201.34 | 100.67 | 50.34 | 134.30 | 67.15 | 302.01 | 151.01 |
| 150 | 400.00 | 200.00 | 100.00 | 50.00 | 133.40 | 66.70 | 300.00 | 150.00 |
| 151 | 397.35 | 198.68 | 99.34 | 49.67 | 132.52 | 66.26 | 298.01 | 149.01 |
| 152 | 394.74 | 197.37 | 98.68 | 49.34 | 131.64 | 65.82 | 296.05 | 148.03 |
| 153 | 392.16 | 196.08 | 98.04 | 49.02 | 130.78 | 65.39 | 294.12 | 147.06 |
| 154 | 389.61 | 194.81 | 97.40 | 48.70 | 129.94 | 64.97 | 292.21 | 146.10 |
| 155 | 387.10 | 193.55 | 96.77 | 48.39 | 129.10 | 64.55 | 290.32 | 145.16 |
| 156 | 384.62 | 192.31 | 96.15 | 48.08 | 128.27 | 64.13 | 288.46 | 144.23 |
| 157 | 382.17 | 191.08 | 95.54 | 47.77 | 127.45 | 63.73 | 286.62 | 143.31 |
| 158 | 379.75 | 189.87 | 94.94 | 47.47 | 126.65 | 63.32 | 284.81 | 142.41 |
| 159 | 377.36 | 188.68 | 94.34 | 47.17 | 125.85 | 62.92 | 283.02 | 141.51 |
| 160 | 375.00 | 187.50 | 93.75 | 46.88 | 125.06 | 62.53 | 281.25 | 140.63 |

# Index